AF588444

# Keeping Chickens

BLOOMSBURY WILDLIFE
Bloomsbury Publishing Plc
50 Bedford Square, London, WC1B 3DP, UK
Bloomsbury Publishing Ireland Limited,
29 Earlsfort Terrace, Dublin 2, D02 AY28, Ireland

BLOOMSBURY, BLOOMSBURY WILDLIFE and the Diana logo are trademarks of
Bloomsbury Publishing Plc

First published in the United Kingdom 2026


A catalogue record for this book is available from the British Library
Library of Congress Cataloguing-in-Publication data has been applied for

ISBN: HB: 978-1-3994-2318-2; ePub: 978-1-3394-2319-9; ePDF: 978-1-3994-2317-5

2 4 6 8 10 9 7 5 3 1

Designed by Austin Taylor
Printed and bound in China by RR Donnelley Asia Printing Solutions Ltd, Dongguan, Guangdong

To find out more about our authors and books visit www.bloomsbury.com and sign up for our newsletters
For product-safety-related questions, contact productsafety@bloomsbury.com

# Keeping Chickens

## How to Choose and Care for Chickens at Home

**Celia Lewis**

BLOOMSBURY WILDLIFE
LONDON • OXFORD • NEW YORK • NEW DELHI • SYDNEY

# List of breeds

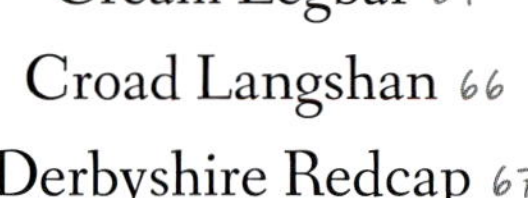

# Contents

# Prologue

**If you've reached** for this book, you've probably already made the decision to join the growing community of those who keep chickens at home. When you're just getting started, learning everything you need to know to care for chickens can feel daunting, but people have kept chickens for meat and eggs for centuries, and whatever you would like to get out of it, looking after chickens in your outdoor space is a wonderfully rewarding experience.

This practical book is easy to use and will provide you with a wealth of information to get you started, including guidance on how to choose the right chickens for you before you buy and preparing your outdoor space before bringing them home, advice on how to feed and take good care of your chickens, and information about breeding them if you decide you'd like to breed your own stock.

Starting on page 47, you'll find detailed profiles for the most commonly kept breeds of chicken, followed by a section with helpful tips on how to use up the eggs you get from your chickens, including some delicious recipes to try out, and inspiring ideas for crafts using eggshells and feathers.

Chickens are a joy to keep. For some, it's about food security and fresh eggs; for others, it's a desire to reconnect with nature or enjoy the companionship of these quirky, intelligent birds. They are adaptable creatures, and whether you keep them in a run or are lucky enough to be able to let them free range in your garden, there will be a breed to suit you. There can be little that compares with the sight and sound of a magnificent cockerel or a mother hen with her 12 little troopers following obediently in her wake. Finding your first newly laid egg will be a thrill, and collecting them daily will remain a pleasure.

Celia Lewis

# A brief history of keeping chickens at home

Which came first: the chicken or the egg? Science has now come up with an answer to this riddle. All new species develop from a genetic mutation, and if the mutation is successful for survival then the new genes are passed on to following generations and a new species is born. The first chicken was a mutation of its avian parents, and its life began in the egg before it hatched. So it was the egg that came first, and it hatched into what eventually became the Red Junglefowl, *Gallus gallus*. The domestic chicken (*Gallus gallus domesticus*) can trace its ancestry to this wild bird, native to the forests of South East Asia.

Domestication of wild fowl likely began more than 8,000 years ago in what is now northern Thailand, southern China and Myanmar. Early humans probably began keeping junglefowl not for meat or eggs, but for cockfighting – a practice that would influence their spread and use for centuries. Cockfights were between two cockerels, whose natural aggression causes them to fight to the death, using either their own natural spurs or, more commonly, by fighting with razor-sharp blades, known as cockspurs or gaffs, attached to their legs. Fights would only end when one bird was killed or was too tired to continue; in many cases, the victor was so severely injured, it also died.

As human societies advanced, chickens spread across Asia and into the Middle East. They reached Europe by at least 800 BCE, brought along ancient trade routes. In the British Isles, chickens were known before the Roman conquest, probably introduced by Iron-Age cultures. When the Romans arrived in Britain, they found chickens already being kept, and cockfighting was well established.

Chickens arrived in the Americas via multiple routes. Spanish explorers brought them to the Caribbean and Central America in the 1500s. Later, European settlers introduced various breeds to North America, where chickens quickly became a farmyard staple.

The earliest uses of chickens for food – eggs and meat – are difficult to precisely date, but archaeological evidence suggests that by 2000 BCE, in parts of the Indus Valley and China, humans were raising chickens with these purposes in mind. Over time, selective breeding enhanced traits like egg production, meat yield and temperament, diversifying the chicken population.

Cockfighting remained a key part of chicken history. Practised in ancient Greece, Rome, India and Persia, it was seen as a test of bravery, status, even spiritual power. The practice persisted for centuries, and European colonists brought it to the Americas. However, growing concern over animal welfare led to bans, first in the United Kingdom in 1835, and later in most other countries, including the United States, where it is now illegal.

Nowadays, there are more than 500 recognised breeds and varieties of domestic chickens all over the world, from commercial hybrids like the White Leghorn to heritage breeds such as the Sussex, Orpington and Rhode Island Red. Sadly, many older or rare breeds, especially those not suited to industrial farming, are in danger of extinction, but some stock does still remain, and it's within anyone's power to help preserve them. Every small flock helps maintain genetic diversity, and keeps old breeds from disappearing. Organisations like the Livestock Conservancy maintain conservation priority lists to promote the preservation of breeds at risk of being lost.

So whether we welcome them into our gardens and backyards for the nostalgia, companionship or the simple joy of collecting newly laid eggs, chickens continue to scratch out a meaningful place in our lives.

Junglefowl cock

Bluebell, a sweet-natured hybrid developed from the Marans and Rhode Island Red breeds

Bovans Nera, a resilient hybrid developed from Barred Plymouth Rock and Rhode Island Red breeds

Braekel, a beautiful proud-looking bird that is also a good forager

# Buying chickens

## What to think about before you buy

If you've decided keeping chickens is for you, before you can get started you will need to work out how much you can afford to spend – on your flock as well as all the equipment you'll need for them – and where your birds will live in your outdoor space. You should also consider what you would like to achieve by keeping chickens at home. Once you know your budget, where you'll house your chickens, and why you want to keep chickens, there are several important questions to ask yourself before you buy.

- Which size of bird is going to be right for you? Chickens can be as small as 15cm (6in) and as large as 66cm (26in). How many birds will you be able to properly care for in the space you have?
- Do you want to keep chickens for the eggs or meat (or both)? Or will your chickens be ornamental and just for show?
- Are you interested in breeding from your chickens? If so you may want to keep pure breeds, which are more expensive but allow for an income to be made from selling chicks. Or, if you're not interested in breeding, hybrid chickens might be a better (and more affordable) option for you.
- If you'd like plenty of fresh eggs and would appreciate an affordable option when you're just starting to keep chickens, could you offer a new home, and a better life, to older hens from battery farms?

Whatever you are hoping to get out of keeping chickens at home, there will be an option that works well for you. The following sections and the information within the breed profiles starting on page 47 will help you to think through everything you need to consider before you buy, and to make the right choices for your circumstances.

### What is a broody?

A broody is a hen with a strong instinct to sit on eggs and incubate them, regardless of whether they are fertilised or not. A broody will typically stop laying eggs when she becomes broody, and will be very protective of the nest, often refusing to leave it except to eat and drink, and becoming agitated if you try to move her. Some breeds, like Silkies and Cochins, are known for their tendency to become broody.

## HOW CHICKENS ARE CATEGORISED

Chicken breeds are primarily classed by their purpose: 'layer' breeds are suitable for laying eggs; breeds that produce a fine carcass are classed as 'table' (or sometimes 'meat' in the US); if breeds are good layers *and* they make a fine carcass, they are classed as 'dual-purpose'; and finally, if breeds are kept purely for show, they are classed as 'ornamental' – these breeds won't be prolific layers and aren't good for eating.

Breeds are also categorised by their size: 'large breeds' are, in fact, standard-size chickens, which are further categorised by their place of origin, so breeds are referred to as American, Asiatic, English, Mediterranean, Continental, etc.; 'bantams' are a miniature version of standard-size fowl.

Nearly all breeds of chickens have a bantam equivalent. There are also several breeds that are naturally small with no large version, which are known as 'true bantams'. Bantams make delightful pets and can become very tame. As you'd expect, they also need much less space than their larger cousins. Most bantam breeds will go broody frequently, making very good mothers, but they do lay very small eggs.

Japanese Bantam hen

Different breeds have different physical traits, and chickens are also categorised by their comb type – there are six different types of comb (see pp. 40–41); by their plumage – a pure breed's plumage will be either 'soft feather' or 'hard feather'; and some breeds have unique characteristics that define their class, including distinct leg feathering, beards and muffs, and the numbers of toes they have.

Soft-feather breeds have loose, fluffy plumage. They come in standard and bantam sizes, and they are predominantly layers, though some soft-feathers are termed dual-purpose if they produce a good meaty carcass as well. The Mediterranean breeds, such as Ancona (p. 49) or Leghorn (pp. 76–77), tend not to go broody and lay good numbers of white eggs. Asian soft-feather breeds are characteristically large with fluffy feathers and feathery legs, such as Brahma (pp. 60–61) and Cochin (pp. 62–63).

Hard-feather breeds have stiffer feathers that appear smooth and sleek. These breeds also come in a range of sizes, including bantam. Gamefowl (such as Asil or Shamo) all have hard, tight feathering, and they're aggressive (as they were developed to fight) and tend to go broody frequently while laying few eggs. This is not to say they are unpopular. Although cockfighting has been banned since the late twentieth century in the UK and early twenty-first century in the US, there is plenty of competition in showing these breeds, and they often have strong and charming characters that endear them to their owners. Keep in mind, however, that they have been bred to be fighting birds, and fight they will. As a result, gamefowl can only be kept in pairs or trios, and no new birds can be introduced as they will simply not be accepted.

From page 49, illustrated profiles of the most commonly kept chicken breeds detail each breed's primary use, size and key physical traits, their tendency to lay or go broody and their unique attributes and personality.

## Chicken or hen?

The word 'chicken' is a general term for a domesticated bird of either sex whereas 'hen' refers solely to females. So, all hens are chickens but not all chickens are hens.

## PURE BREED OR HYBRID CHICKENS

If you don't want to breed from your chickens, hybrids can be an economical way of starting your flock. Hybrids comprise two or more pure breeds that have been carefully selected to produce birds with prolific egg-laying tendencies, hardiness, disease resistance and good temperament. There are many well-known hybrids, and there will likely be a breeder near you that stocks one or more of them. Hybrids are the most economical egg-producing machines you can purchase, so if you don't want chicks or a cockerel, hybrids are for you as they very rarely go broody and their chicks may not inherit the traits of the parent hybrid. So, although a parent hybrid may be a good layer or have a good temperament, this is not guaranteed in the offspring.

Speckledy, an easy-going hybrid developed from Marans and Rhode Island Red in the UK

## Other factors you must consider before you buy

- **Do you want your hens to lay fertile eggs so you can breed?** If so, you will also need to buy a cock (known as a cockerel until his first moult). See pages 31–37 for more on breeding.
- **Will you or someone else be able to feed your flock twice a day?** Do you have neighbours or friends who can step in when you are away from home? Automatic feeders are available to purchase, but in cold weather water will freeze, and hens drink a surprising amount. See pages 25–26 for more on chickens' food and water needs.
- **Do you have enough space for a henhouse and run for the number and type of chickens you want to keep?** Some breeds need more space than others. See pages 19–21 for more on housing needs.
- **Your henhouse and run will need regular cleaning,** and remember that where there are hens, there may also be rats. See pages 27–28 for more on care and hygiene.
- **Your chickens' health will need to be checked regularly,** and you may need to treat them for pests like fleas or red spider mite. See pages 29–30 for more on pests and diseases.
- **Are there predators in your vicinity?** If there are, you will need a high, strong fence to protect your chickens. See pages 22–24 for information on how to keep your flock safe.
- **Are you capable of administering a coup de grâce,** if necessary? See page 29 on culling.
- **Local regulations.** Check the regulations for keeping hens, as they may differ in different areas or countries.
- **Flock size.** Chickens are flock animals and will not be happy living on their own. Keep at least three to ensure that they are content.

Pure breeds won't lay quite as many eggs as hybrids, and some will tend to go broody, but you will be able to breed from them. Pure breeds reliably pass on their traits to their chicks, following breed standards. They come in an amazing array of colours, characters, shapes and sizes. Some pure breeds are better layers than others, some produce better carcasses and some just look extraordinary. Certain pure breeds are rare, and it can be very rewarding to keep a rare breed and help preserve it.

## RESCUE HENS

Rehoming commercial hens is another affordable option to consider when you are starting your flock. Commercial farmers find it most profitable to have a constant supply of young birds that are coming into lay (beginning to lay their first eggs), and farmers often cull hens when they reach 18 months old to make room for younger birds.

There are several charities that try to help rehome these older birds instead, including the British Hen Welfare Trust and Fresh Start for Hens in the UK. In the US, the Adopt a Bird Network and the Farm Animal Adoption Network rehome rescued hens and roosters. Charities will expect to see a photo of where the hens are going to be rehomed and to receive a small donation for each bird.

These hens may be termed free-range, although it is likely they have been kept in crowded conditions and may be a bit battered and lacking feathers when they arrive. However, they will very soon perk up, regain their feathers and continue to lay. Being commercial hybrids, they lay well but may not be as long lived as pure breeds. At first, they may even be too scared to come out of the henhouse, so it is wise to let them have food and water inside for a day or two until they venture outside.

Lohmann Brown, a popular hybrid developed in Germany from the Rhode Island and New Hampshire breeds

## Life stages of a hen

Life expectancy of chickens depends on the breed, care and environment but is typically between five and ten years. Egg production tends to decrease as a chicken gets older and can stop entirely as they approach the end of their life.

**Fertilised egg:** this stage, from when the egg is laid to hatching, takes around 22 days.

**Chick:** from hatching, the chick stage lasts around 65 days.

**Pullet:** a young female chicken in her first year of laying. This stage is the equivalent of adolescence in humans. Pullets will begin laying eggs from 15 weeks old.

**Hen:** a hen is an adult female chicken, at least one year old and generally capable of laying eggs.

# Where to buy your chickens

When you've decided on your breed, you will now need to acquire some birds.

If you want to start from scratch, you will need fertile eggs. These can come quite safely by post, and you can find breeders with hatching eggs for sale by looking in poultry magazines or online, although make sure they are reputable breeders and not just selling you supermarket eggs. Another good source would be poultry club or agricultural shows, and some country markets also have large poultry sections with live birds, as well as hatching eggs.

To hatch the eggs, you will either need to borrow a broody hen or buy a small incubator. Eggs that come by post should be given at least a 12-hour rest period before being introduced to the broody or incubator. Be aware that some of the eggs may not hatch, and certainly some – possibly half – will be cockerels. For more on hatching eggs, see pages 36–37.

If you want to head straight into adult birds or chicks, you will have to find them closer to home – you can find lists of reputable breeders in poultry magazines or approach a particular breed club or local poultry society.

Day-old chicks are an option, and although some breeds can be sexed at hatching you may still find you have a plethora of cockerels that must either be given away or culled. The younger the bird, the less it costs; chicks are 'off heat' at about five weeks, meaning they no longer need their mother or a heat lamp, although this may still be too early to tell which are pullets and which are cockerels.

The best option is to buy birds that are 'point of lay', which is traditionally when they are between 18 and 21 weeks old. However, they will be considerably more expensive, and if you buy them in the autumn, there is no guarantee that they will start laying until the days start to lengthen in the spring.

As mentioned on page 15, another affordable option is giving a home to rescue hens, which will already be in lay and will cost very little. Although they may be scruffy and nervous at first, given a good home, your rescue hens will soon settle in.

### Legal requirements for keeping chickens at home

**In the UK:** even if you have only one chicken, you must register it with the Animal and Plant Health Agency (APHA). You are also expected to adhere to welfare standards and ensure that you are not causing a nuisance to neighbours.

**In the US:** regulations for keeping chickens vary significantly by state, county and even municipality, so it's crucial to check local ordinances before acquiring any bird. Many municipalities have specific rules regarding permits, number of chickens allowed, coop size, and whether roosters are permitted. Some areas may even have restrictions on free-ranging chickens.

## Bringing your flock home

Hens are totally docile in the dark, and easy to pick up and move, so having collected your birds, let them stay in their boxes until dark and then introduce them to their henhouse. If these are your first birds and they have no older ones to follow, they may not know to go into their henhouse at night – you will have to teach them. Each night, check if they have gone into their henhouse and if not, retrieve them from their chosen spot and set them on the perch; they will very soon get the message.

# Housing your chickens

**Hens are hardy creatures,** but they need a degree of pampering if they are going to lay well. You will need a henhouse (or coop) of some sort even if you want your chickens to free range during the day (see Keeping free-range chickens, p.20).

All kinds of purpose-built henhouses are available in a wide range of sizes, but before you can work out which is right for your space and your chickens, you need to decide how large your flock is going to be. Each hen will require about 25cm (10in) of perching space, and there should be a nesting box per four or five birds.

Purpose-built henhouses come with ready-made pop-holes, nesting boxes and perches, but can be expensive. A less expensive option is to buy the smallest tool shed available and convert it into a henhouse by making a pop-hole (25 × 30cm / 10 × 12in for average-size hens), adding a perch and putting in some nesting boxes, which could simply be wooden wine cases or even cardboard boxes filled with either hay or wood shavings.

If you will be keeping bantams, another alternative is a fold unit or ark. They are very versatile housing with a small run attached and can be moved around with ease, giving access to fresh grass. Fold units are small, so bantams are the most suitable occupants, and even a large ark is very cramped for a full-size hen.

A tool shed that has been converted into a henhouse

Moveable fold unit

## Where to site the henhouse and run

Firstly, remember you will be visiting your run twice a day, so you do not want it too far away.

You will also need to ensure that your chosen site will offer your chickens a mixture of shade and access to sun, and that it has a sheltered area to keep them out of the wind and rain. Ensuring the site has a permanently dry area for feeding is particularly important, and the hens will naturally use it to shelter from rain. Providing them with a box of sand or wood ash to dust bathe in is advisable; if you don't, they'll create patches of loose soil on the ground themselves.

Do you have close neighbours? A crowing cockerel may be your idea of a pleasant morning alarm call but you can't presume that everyone nearby will share your view. Hens also make quite a lot of noise when they lay an egg, and this too can be an irritation to anyone living close by – though they may be less inclined to mind if they're presented with the occasional box of freshly laid eggs.

## Keeping free-range chickens

If you would like to keep free-range chickens, to keep them safe at night you will have to shut them in their henhouse each night and let them out again in the morning – no more sleeping in on a Sunday. If you don't shut them in when they are most at risk from predators, you may not initially experience a full-scale slaughter, but once a fox or other predator has come across your free-ranging chickens it will return again and again until none are left.

Your chickens will go to bed at dusk, which, depending on where you live, in summer could be as late as 9:30 p.m. They will learn to put themselves to bed and will naturally get up again as soon as it is fully light outside, but this may be 5 a.m. in summer, when predators still abound. Nowadays, you can acquire an automatic pop-hole door. This is a small battery-powered gadget that opens and closes the pop-hole depending on light sensitivity. It can also be set to activate at specific times if required, but you need to remember that bedtime will alter with the daylight and disaster could ensue if the timer closes the door of the henhouse before your free-range chickens have entered.

A sensible precaution against predators is to have a small fox-proof run around the henhouse so that the hens can be let out to freely range during the day but be safely shut in when they have their corn within their safe area in the evening, and they can go to bed and get up when they like.

Free-range hens will be utterly content, but if you give them complete access to your outdoor space they will scratch around and have dust baths in your flower beds and munch merrily through your lettuces and cabbages if they have access to your vegetable plots.

An additional problem with free-ranging hens is that they may develop the habit of laying out, making their eggs difficult to find. They will search out a hidden spot and be secretive about using it. Undercover detective work will be required to spot the hen as she slinks off, and once her preferred laying spot is discovered, she will find another hidden place. Hens mainly lay in the morning, so if it fits with your day, letting them out of the henhouse a bit later can help to prevent free-range hens from laying out.

Most likely, you will have to keep them in a run – as large as you can make it – and unless it is extremely large, it will eventually become bare from their constant scratching. The hens, however, will be completely happy and you can throw in weeds, gone-to-seed vegetables or even straw for them to forage about in. If space is no problem, make two runs side by side that can both be accessed from the henhouse and that way one can be rested while the other is in use.

Your free-range chickens will also need a daytime shelter of some sort to shelter from rain or sun, a dry spot for dust bathing and somewhere for them to be fed when it's wet. One idea to do this is to put your henhouse up on blocks so that there is a good dry area underneath, with a ramp for the hens to walk up to enter the henhouse.

A henhouse on legs with convenient shelter space underneath

# Keeping chickens safe

## Pecking order

Flocks of hens soon establish a pecking order, and any hens that are introduced to the flock will be at the bottom of it so, to prevent bullying, never introduce only one hen at a time – always add two or more new arrivals together. Hens can be very unkind to newcomers at first, pecking and chasing them off the food, but the new hens will soon settle in. The exception is gamefowl, which are so aggressive that newcomers will never be accepted, so these breeds can only be kept in very small groups, possibly just a pair of hens, or a 'trio', which is one male bird and two females.

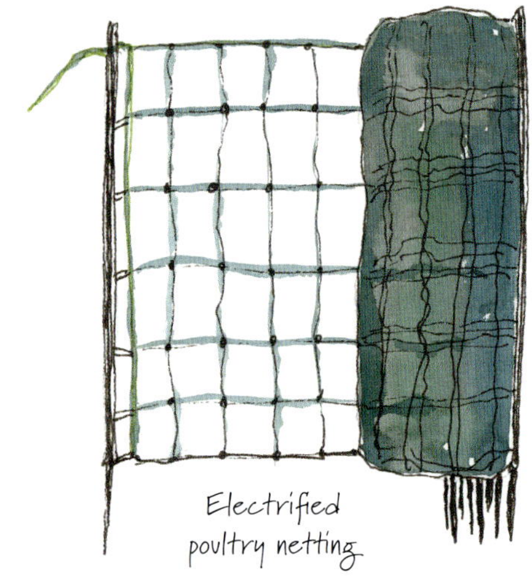
Electrified poultry netting

## Precautions to protect your flock

### SECURING RUNS FROM PREDATORS

Unless your hens are going to be completely free-range, you will need a fox-proof run. To be fox-proof, your run should have a 2.1–2.4m (7–8ft) high chicken-wire fence that is either dug 45cm (18in) into the ground or that has 45cm (18in) of wire netting laid flat on the ground on the outside of the fence to prevent foxes digging in and under the fence.

It's sensible to add some electric wires to the fencing around your run; even just a couple of strands at the bottom and one at the top will help to deter predators. Electrified poultry netting is also available, which has the advantage of being easy to move to a clean patch of ground. You will require a battery-, mains- or solar-powered electrifier.

Some henhouses come complete with a run attached. These are very neat, requiring little space, but they may not be predator-proof. If the sides of the attached run simply rest on the ground, this leaves your flock vulnerable.

Foxes are a formidable and persistent foe. If they manage to get into your run, they will kill all the hens, whether they can carry them away or not.

Your hens don't only need protection from foxes. Some birds of prey can carry off a fully-grown hen, including red kites and buzzards, populations of which are increasing in the British Isles. If they can get to

Henhouse with a small run and shelter

them, smaller hawks will also take chicks and young birds. Hanging netting and strings that have something sparkly on them across your run will prevent this. Birds of prey will be unable to access the coop from above and are deterred by flashes of light.

### WING CLIPPING

Even when you have predator-proof fencing around your run, some of the more flighty or lighter breeds may decide to fly over the fence. If this happens, clipping one wing is the solution. It's a completely painless procedure for the hen – on a par with trimming our nails.

Clipping a wing will unbalance the bird and make it difficult for her to fly. It is important that you only clip one wing, or the bird will not be unbalanced.

Wing clipping a hen will be easier if you have a helper. Ask your helper to hold your hen, then take one of her wings in your hand and, using scissors, cut halfway down across the primary feathers. The primary feathers will be the first few (usually 10) long feathers at the front of the wing (see illustration on p. 42). You will have to repeat the procedure after each moult, as the feathers will grow back.

## Identifying predators

Unfortunately, chickens, even those that aren't free-range, have many predators, including dogs, foxes and badgers and, in the US, also coyotes and even snakes. If your flock is attacked, you need to be able to recognise who the predator was and review the precautions you have in place to try and prevent further losses.

When predators attack your chickens, they always leave evidence behind. Here are the clues to look for and pointers to help you deduce who the likely perpetrator was:

- Were the hens in their run, and if so, how did the predator gain entry?
- Was more than one bird killed?
- Were the birds removed?
- Were any parts of the birds left behind?

| Evidence | Perpetrator | Preventing future attacks |
|---|---|---|
| **Many chickens injured or killed but not eaten.** | Dog | Report the attack to the police – dog owners are responsible for paying for any damage caused.<br><br>Ensure fencing extends below the ground (45cm/18in) so the animal cannot dig their way in.<br><br>Secure fencing and coop. |
| **Most chickens killed. Heads bitten off. Feathers everywhere. Some birds partially buried. Some birds removed. Hole torn in fence or feathers left where the predator climbed over and jumped down while carrying one away.** | Fox or coyote | Ensure fencing extends below the ground (45cm/18in) so the animal cannot dig their way in.<br><br>Try electric fencing. |
| **Battered housing, many birds killed, some eaten but none removed.** | American or European badger | Secure housing.<br><br>Electric fencing. |
| **Birds going missing one at a time over several days with no sign of entry.** | Bird of prey | Roof over chicken run. |
| **Bodies left in run but heads removed.** | Owl | Roof over chicken run. |
| | Racoon | Secure latches (as raccoons can open doors). Ensure fencing extends below the ground (45cm/18in) so the animal cannot dig their way in.<br><br>Fill any holes or gaps in the coop's walls. Use a tight wire mesh to reinforce the chicken's enclosure and coop, rather than chicken wire. |
| **Remains of birds left in run – abdomen eaten.** | Opossum | Use a tight wire mesh to reinforce the chickens' enclosure and coop, rather than chicken wire. Consider installing lighting as a deterrent. |

When you work out what attacked your flock, always review the protections you have in place, because once a predator finds your chickens, it will keep coming back until it gets them all. Use your findings to reinforce existing or add more protection around their henhouse to keep the rest of your chickens safe.

# Feeding your chickens

## What to feed?

There are many and varied views on how to feed your chickens. Your local feed merchant will stock mixed poultry corn, chick crumbs, growers' pellets, layers' mash and layers' pellets. Your hens can eat anything you can eat, as long as it has been cooked, as well as uncooked things like lettuce or cabbage leaves from your vegetable garden, and feeding your chickens greens helps to make their yolks orange. They will also enjoy potato and other vegetable peelings, but these must be boiled up for them. They will thoroughly enjoy dried bread crusts, cake and stale biscuits, but avoid anything salty, as birds have an intolerance to salt. They can't eat such things as banana skin, orange peel or tea bags.

In the UK, it is illegal to feed hens any food from a domestic or commercial kitchen, even from a vegan household.

### WHAT ELSE TO PROVIDE WITH THEIR FOOD?

You will also need to supply your chickens with oyster shell via a separate feeder – to replace the calcium needed to make eggshells – and mix some grit with their feed, which they use to grind up their food in their gizzards to aid digestion.

## How much to feed?

This is never an exact science. It all depends on the size and age of your hens, how much food they are going to be able to find for themselves, what the weather is like, what quantity of feed you have, etc. As a very simple rule of thumb, a large handful of pellets and a large handful of corn for each bird per feed is about right. Feeding is something you will eventually get a feel for, and if you notice there is anything left after an hour, you are probably feeding them too much and should cut back, as any food left lying around for long will encourage rats and other vermin. If the feed you provide has all gone in 15 minutes, this is probably too little, and you should up their portions until you reach the optimum amount for your chickens.

The following is purely a guide and presumes that your hens do not have access to any other food.

- A chick of six weeks old will require approximately 50g (2oz) per day, divided between chick crumbs and corn.
- A grower of 12 weeks will require approximately 75g (3oz) per day, divided between growers' or layers' pellets, or mash and corn.
- A laying hen will require 125g (4.5oz) per day of corn and either pellets or mash.
- You could always just put corn and mash/pellets in hoppers and let them help themselves ad libitum, which would be fine if you were away for a day or two, but using this as a permanent means of feeding can lead to the hens getting over-fat and lazy.

## When to feed?

The best advice on when to feed your chickens is to find a routine that fits in with your own. Hens are versatile creatures, and you can put in as much or as little time as you like with your chickens; however, if you completely automate their feeding, you will miss half the fun of caring for a flock.

Ideally, give the mash or pellets feed to them first thing in the morning and the corn an hour or so before bed, or in late afternoon in the summer – this is also the time to collect any eggs. Feeding time is the ideal moment to get to know your hens and develop an eye for trouble before it occurs. Does one look out of sorts? Is another being bullied? If you're familiar with your birds, you will minimise the risk of problems.

## Storing your feed

Metal dustbin

All food types suitable for chickens are perishable and must be kept dry. As the feed ages, it will degrade in quality and may become infested with insects or mould, so it makes sense not to buy more than you are going to use over a few weeks.

Metal dustbins are ideal for storing your feed. Plastic containers may very soon be chewed through by squirrels, who are particularly partial to hen feed (and a free meal). Obsolete freezers make excellent airtight storage containers.

If you prefer, you can leave the feed in the bag when you store it, which will ensure that bag is totally finished before you start the next. If you are going to pour the grain or pellets into a bin, make sure the bin is empty first, tipping anything that remains into a separate container. Then fill the bin with the new feed and place the old feed on top. This ensures a proper turnover.

## Providing water

Your hens will drink a surprising amount of water, especially when they are laying, so make sure they always have a plentiful supply, and if their water supply freezes in winter, ensure you defrost it for them daily. This is especially important if you will be away for a day or two when there is any chance of frost – you must get someone to come in and check every day that your water supply hasn't frozen solid. Specially designed water towers are best, but small buckets will also suffice as long as the hen can reach the water. Narrow-lipped drinkers will be required for birds with fancy feathering on their heads, to keep their feathers from getting wet.

Water tower

Ad lib feeder

# Caring for your chickens

### DAILY CARE

Every day, you must:

- Let the hens out if they are confined to their henhouse.
- Feed mash or pellets first thing in the morning.
- Check their water and refill it, if necessary.
- De-ice their water containers in frosty weather.
- Cast your eye over your flock – do this daily and you will immediately spot if there is anything amiss.
- Feed grain in the late afternoon in summer, otherwise an hour before they go to bed.
- Collect the eggs.
- Shut the birds up at dusk.

### WEEKLY CARE

Every week, you need to:

- Clean the henhouse and refresh the nest box bedding, changing or topping up the contents as needed.
- Top up the oyster shell and grit.
- Look out for rats and other vermin and take precautions.
- Check all fencing is secure.

### YEARLY CARE

Throughout the year, you should:

- Think about breeding.
- Consider fixing leg rings to birds you wish to keep an eye on (especially if you have a lot of chickens of the same breed).
- Worm your birds in spring and autumn.
- Clip wings of flighty birds after they have moulted (see p. 23).
- If you have a moveable henhouse, move it to fresh ground regularly.
- If you have a fixed henhouse, check its weatherproofing.
- Review your flock – are you getting enough eggs?

Metal water tower

# Maintaining good hygiene

Periodically, you will need to clean out your henhouse. If it is not regularly cleaned, your flock may get red spider mite, fleas or lice, particularly if they live in crowded conditions. There are plenty of preventative products on the market, some of which are organic, and if you sprinkle the powder or spray around the henhouse, particularly on perches and in nesting boxes, each time you clean out there should be no problem. It is also a good idea to change the contents of the nesting boxes frequently. When you're cleaning out the henhouse, remember that chicken droppings are high in nitrogen and make excellent fertiliser, or you can just add them to your compost heap.

# Storing your eggs

Eggs should be stored in a cool, dry place – ideally not in the fridge. They should be kept in containers, otherwise they can absorb the smell of nearby food. Eggshells are very porous.

Egg cartons, recycled from supermarkets or bought, keep eggs fresh and safe. You can mark the collection date on the egg with a pencil to keep track of the eggs and make sure you use up the oldest eggs first. Other containers like an 'egg skelter' can also help you keep eggs in order.

If you have too many eggs, you can freeze them in a freezer-safe muffin container. Remove the shell before freezing as bacteria can contaminate the insides as they expand and crack the shell. Lightly beat the whites and yolk together and add a pinch of salt or sugar before freezing. The salt or sugar stops the yolk becoming gelatinous. Once frozen, use within a few months. When you want to use the eggs, let them fully thaw and use straight away.

Always wash and dry your hands before and after handling eggs, as some eggs can contain salmonella bacteria inside or on their shells.

# Egg sizes and colours

Egg sizes and colours vary between breeds, with some laying large eggs and some small; some breeds even lay blue eggs. A pullet's first eggs (known as witch's eggs) will be small, tiny even, with no yolk, but the eggs will gradually increase in size, and as the hen ages, her eggs become fewer but larger.

# Egg issues

Occasionally, you will find an egg with a soft shell – this may occur when the hen is starting to moult or beginning to slow down on production.

Very thin, brittle shells may signify a lack of calcium. The hen may need more oyster shell. Or she could be laying so many eggs that she is unable to produce enough calcium to provide shells for all her eggs.

Sometimes a shell is wrinkled – this is generally a problem for older birds whose oviduct may be worn. As long as the shell is not broken, soft or wrinkled eggs can still be used as the contents will be identical to normal eggs.

Every so often, you will find an egg that has two yolks – this usually occurs when a hen is coming back into lay after a rest period. Blood or meat spots in the yolk are common, and although they look unattractive they are harmless.

Some hens, especially pure breeds, will slow down their egg laying in the autumn when the daylight hours are shorter, if not stop altogether. Once the days lengthen, laying should start again. Hens also stop laying while they moult. Stress can also be a factor, such as that caused by a move from one place to another or the addition of an unfamiliar hen to the flock. Another factor for decreasing egg production is age – the older the hen, the less she will lay.

## Culling

There will eventually come a time when one of your flock may need to be put out of its misery – possibly it will have been attacked by a dog or become seriously ill – and it is your responsibility to do this. If you hatch your own chicks, you must be aware that 50 per cent will most likely be cockerels, which you may not even be able to give away, let alone sell. The obvious answer is to rear them for the table and end their lives in the most humane way possible.

You could take them to the vet or find another poultry keeper nearby who will do the job for you – your local poultry club could help with this. There are various humane dispatchers available, but it is generally thought, and recommended by the Humane Slaughter Association, that neck dislocation is the kindest and swiftest method. Before attempting it yourself, it would be a good idea to find someone to give you a demonstration.

### HOW TO WRING A NECK

If possible, collect the bird for slaughter after dark when it is quietly on its perch. Hold the legs firmly in one hand, and with the bird's body against your thigh and head hanging down, place the index and middle fingers (or thumb and index finger, if you prefer) on either side of the neck with the head under your hand. Bend the head slightly outward and give a sharp, twisting pull downwards, at the same time bending the neck backwards. This dislocates the neck from the head and death follows instantly. There will almost always be a certain amount of flapping but this is simply post-mortal nerves and will cease shortly.

## Common pests and diseases

| | Problem | Signs | Action |
|---|---|---|---|
| Parasites | **Fleas** | Birds may scratch and create bare patches – poor laying – look unwell. | Treat with flea powder, including in henhouse. Increase hygiene in henhouse. |
| | **Lice** | Similar to above – check vent for white clusters of eggs. | As above. |
| | **Northern fowl mite** | Dirty-looking feathers on pale-coloured birds. This mite lives on the bird. | Increase hygiene in henhouse. Flea powder or proprietary spray every 8 days. |
| | **Red spider mite** | This mite lives in the henhouse, on perches, etc. – you will be itchy if you have been inside. | As for above but cycle is 7–10 days. |
| | **Worms** | Hard to spot. | Worm birds twice a year with a proprietary product. |
| | **Scaly leg** | Scales lifted on the leg – leg looks crusty and enlarged because of the tiny mite that burrows under the scales. Legs return to normal after next moult, when cured. | Proprietary anti-scaly leg product or immerse leg for 20 seconds in a jar of surgical spirit once a week for 4 weeks. |

*Continues overleaf*

# Common pests and diseases, *continued*

| | Problem | Signs | Action |
|---|---|---|---|
| Illnesses | Crop bound | Sour-crop – bulging soft crop when bird gets up in morning – hen appears ill – caused by fungus called *Candida albicans*.<br><br>Impacted crop – full, hard crop when bird gets up – can be caused by eating long grass. | Feed live yoghurt – encourage vomiting by holding bird's head down and massaging crop.<br><br>As above – gently massage crop. |
| | Coccidiosis | Milky white diarrhoea, occasionally with blood. 3–8 weeks old danger period. Sudden death. | Hatch as early as possible in the year – parasite lives on damp grass, spread in droppings.<br><br>Good hygiene. Vaccine available. Chick crumbs and pellets contain anti-coccidiostats. |
| | Colds | Runny nose and coughing. | Proprietary preparation from vet or feed merchant. |
| | Marek's disease | Many and varied – dead birds, paralysis, not growing, unnaturally hungry. | Vaccination at day old. Never mix young with old birds. |
| | Newcastle disease (fowl pest) | Notifiable. Twisted necks, birds floppy and unsteady, breathing difficulty. Death. | Vaccination – if diagnosed, mandatory slaughter of flock. |
| | Avian influenza (bird flu) | H5N1 breeds in respiratory and intestinal tracts and is transmitted bird to bird.<br><br>Difficulty breathing and sudden death. | Hygiene – covering run to keep out wild birds during outbreak.<br><br>In the UK, cases must be reported immediately to APHA. |
| Other problems | Fishy eggs | Eggs have fishy flavour. | Found occasionally, no known cause or cure. |
| | Egg eating | Birds start eating eggs in nesting boxes. | Starts with broken egg – keep supply of shell available. Collect eggs frequently. Make nesting boxes dark. |
| | Feather pecking | Hens develop bare, sore patches on neck, rump and vent. | Less overcrowding – outdoor perches and dust baths to combat boredom. |

# Breeding your own stock

## Cocks or cockerels

You do not need a cock (known as a cockerel until his first moult) for your hens to lay eggs; they will do this anyway, but if you don't have a cock, your hens' eggs will not be fertile.

There are definite pros and cons to keeping a cock. On the pro side, you will be able to hatch chicks, he will look magnificent and, to an extent, he will protect his flock from predators or at least be the first line of defence. He will also be charming to the hens, calling them over when he finds some food – he has an ulterior motive, of course, but the hens always respond.

On the con side, he will crow. He will crow frequently during the day and start before dawn – various bantam cocks even crow in the middle of the night. To some, this will be a delightful sound of the country, but to others it will be an unpleasant nuisance, so this should be an important consideration. You can keep him quieter by making sure he stays in a dark henhouse until a reasonable hour; he will also find it difficult to crow if he can't lift up his head, so a high perch or cramped conditions may keep him quiet to an extent.

If you only have three or four hens his favourite one may suffer from his attentions and develop a bare back and neck. Although unsightly, this will not harm the hen. However, if you want to show your birds, you can acquire a 'saddle', which is a sort of light cover that fits on the hen's back. A cock can service at least 10 hens, so keeping six or more should spread his attentions.

Finally, he may very well be aggressive. He will certainly be aggressive towards another cock but he may also find you a threat and take, literally, to attacking the hand that feeds him. However, plenty of cocks are perfectly friendly, particularly the larger breeds, and even gamefowl tend only to be aggressive towards their own kind.

Don't forget he will be another mouth to feed with no return – unless you are considering coq au vin.

A broody hen with her clutch of eggs

### COCKS AND FERTILITY

When a cock is introduced to the flock, the laying hens will become fertile after a couple of days. To be sure of selecting fertile eggs, you need to choose ones which have been laid at least a week afterwards. If your cock should suddenly die for whatever reason, the hens' eggs will remain fertile for up to four weeks – it therefore follows that if you have two cocks running with your flock but only want one to fertilise the hens, it will be four weeks from the day you remove the unwanted cock until you can be certain all the progeny will be the remaining cock's.

In the poultry world inbreeding is normal; in fact, many distinct strains are created by what is called 'line-breeding', where the entire strain emanates from one cock and hen. As long as the best layers and lookers are used, there should be no problem, although eventually inbreeding will cause infertility, so occasional new blood should be introduced.

## Broody hens

A hen is described as 'broody' when she feels she has laid her clutch and is ready to start sitting on or incubating them. You will find she has remained in the nest box all day and night and fluffs up her feathers if you put your hand under her to remove eggs. She may peck your hand and will refuse to get off the nest.

### PREVENTING BROODINESS

Your hens may still go broody even if you don't have a cock, but their eggs will not be fertile, so you must decide whether to acquire some fertile hatching eggs from somewhere else or try and dissuade broodiness in your hens. Dissuading a broody hen will not be easy, but what she wants is to think she is sitting on her clutch, so preventing her from getting to the nest box is the first move. If you can, put her outside the run (she will be easy to pick up) for a spell each day; she will run up and down trying to get back in and eventually forget her broodiness, but it may be some time before she starts to lay again. Another method is to have a small henhouse with a slatted or wire-netting floor, where any breeze will cool her breast and she will be unable to sit. One of the strange contradictions of poultry-keeping is that when you don't need a broody, a hen that goes broody will be very difficult to discourage from her nest, but when you want to hatch eggs and have been waiting patiently for one of your hens to go broody, when you try to move her onto a new nest, if not handled carefully, she may stop her sitting at once!

## REARING CHICKS WITH A BROODY

If you want to breed, having a broody is a joy as she will do the work of hatching the eggs for you. Before she can do this, you will need to make sure she really is broody. If possible, move her (tactfully, at night) to a nest box well away from the other hens. If she remains with the other hens, she will allow them to add daily to her clutch, and you will have to try and sort out the newly laid eggs as any of the new (fertilised eggs) will obviously have a different hatching date. So, move her away from the others, and set her on some old or china eggs, even golf balls would do, to make sure she is settled.

You have a week or so to collect your own fertile eggs or acquire some. Choose clean, well-shaped eggs and keep them in a cool place; they should not be warm, about 12°C (55°F) is fine. Do not wash or wipe them but place them pointed end down in an egg box and alter their position once a day. An easy way to do this is to prop the box up on one side and swap sides each day. They will keep for at least 14 days, but hatchability goes down after about 10 days. Eggs that have come by post should be no more than seven days old, and these should be given at least 12 hours to rest before being introduced to the broody.

Once you have collected your clutch, which can be up to 12 eggs for a full-size hen, or 12 bantam's eggs or five or six full-size eggs for a bantam, remove the false eggs and pop the hatching eggs under the broody, being careful not to upset her – this is best done at night.

She will now settle in for the 21 days it takes for the eggs to hatch, carefully turning them herself several times a day. She will probably only leave her nest once a day to drink, eat and defecate – do not worry if you do not see her get off and try not to disturb her. Leave her food nearby– she will be happy with a couple of handfuls of wheat – and make sure she has fresh water.

In her natural state, when the hen gets off her nest to eat and drink, she would dampen her breast in the dew on the grass. If she is confined to a shed or has no access to outside, it is a good idea to sprinkle the eggs with lukewarm water very carefully on the last few days before hatching is due – chicks formed but dead in the shell when the hatching date is reached is a sign of lack of humidity.

A broody hen keeping her chicks warm

On the 21st day, the eggs will start to 'pip', which is when the chick, using its egg tooth (a horny growth on the end of its beak), begins to break through the shell. Once it has made a small hole, it will rest for up to eight hours; at this point you can often hear the chick already cheeping from inside its shell. It then continues the laborious task of breaking free. The mother will wait until she is sure that all the chicks have hatched and dried before bringing them out – chicks can happily survive for 36 hours with no food or water, living off the remaining yolk in their stomachs. If, after 36 hours, you suspect there are still unhatched eggs, it is best to remove them. Have ready a low container of chick crumbs and a water container that the chicks can reach into but not drown in – the addition of a few stones will help to make it safe. Don't forget food for the broody herself; you can give her mixed corn, and she will break it up for the chicks and show and encourage them to eat the crumbs.

# Using an incubator

Without a broody, you can still rear chicks by using an incubator – the results are never quite as good, and you should expect the occasional failure. Incubators come in all sorts and sizes, from tiny models that will require hand turning of the eggs several times a day, to medium-sized that will take 20–40 eggs and do the turning automatically, to enormous commercial cabinets that may take several hundred eggs. There are two types of incubators: forced-air machines that have a fan built in to circulate the air and still-air incubators with no fan. What they all have in common is that the humidity and temperature must be correct: 37.7°C (100°F) for forced air and 39.4°C (103°F) for still air. The relative humidity should generally be 45 to 55 percent, increasing to 65 to 75 percent in the last few days before hatching. Humidity levels can be measured by a hygrometer or wet-dry bulb.

A medium-sized incubator that can house 20–40 eggs

Machines will all be slightly different, so read the instructions that come with your machine to find out exactly what temperature should be set and how much and where the water for humidity should be added. During incubation, the water will evaporate and must be topped up – always add warm water, as close as possible to the temperature in the incubator. Each machine will have a ventilation hole – follow instructions for how open or closed it should be.

## Candling

After eight days, you can 'candle' the eggs to see if they are fertile. This involves holding each egg in front of a bright light. You can buy special candling lamps but home-made versions work just as well. You need a bulb or torch inside a box with an egg-shaped hole on the top. The egg is held over the hole so that the light shines through and you can see inside. This is best done in a darkened room. At seven to eight days, a fertile egg will have blood vessels that look rather like leggy spiders and an obvious air sac at the broad end. At 14 days, any infertile eggs will appear clear and should be removed, and those growing correctly will have a large shadow and enlarged air sac.

You will be able to see the embryo much more easily through a white shell – eggs from Marans or Welsummers have exceptionally dark shells which makes candling much more difficult, and you may not be able to see anything at all through them.

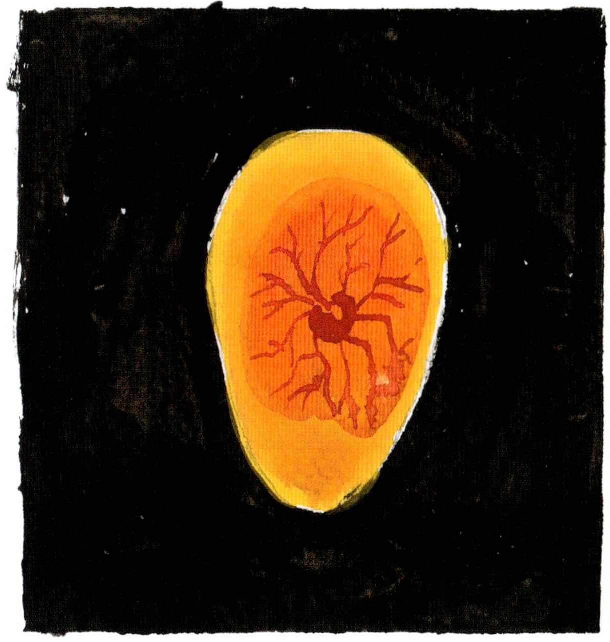

# Egg hatching

For the last three days before hatching, remove the machine from its cradle, or stop turning, and increase the humidity slightly. To maintain humidity, try to resist opening the machine when hatching starts (this will be hard!). Pipping will start on the 21st day, and even before that you may hear cheeping. As you approach this day, remember that you'll need to have a rearing pen with a heat lamp ready to place your newly hatched chicks into when they are dried and fluffed up.

The chick will break through the shell with its special egg tooth. Remember that once pipped the chick may rest for at least eight hours before continuing to hatch, and the whole process may take 24 hours. Never help a chick from its shell – it will hatch on its own or there will be a reason why it doesn't. At first, the chick will look a little bedraggled, but it will very soon fluff up.

If you are unlucky enough to have a power cut while incubating eggs, all may not be lost. There are various methods of keeping the eggs warm. You can cover the incubator with blankets, or find a box large enough to fit over it which will conserve the heat. Alternatively, make a makeshift frame that fits over the incubator and light nightlight candles in jam jars inside the frame. This should be sufficient to maintain warmth until the power returns. Embryos can survive at slightly lower temperatures for up to 18 hours, although this may mean the hatch will be one day later.

The chicks will rest for a while when they have finally broken free of their shell, but they'll soon be on their feet. As soon as chicks have completely dried and fluffed up, remove them to the rearing pen with heat lamp that you have already prepared.

If you have room, create a circular enclosure with cardboard or other flexible material so that there are no corners for chicks to get trapped in. Hang the heat lamp slightly to one side and at a height that ensures the temperature is 32°C (90°F) at ground level.

The chicks will also need shallow containers for chick crumbs and special chick drinkers. Chick crumbs are a specially formulated food created from grains, vitamins and minerals that are crushed into small crumbs that are easily digested by the chicks. Chicks should be fed ad lib so they can help themselves.

When you are ready to introduce chicks to their new home, dip their beaks carefully in the water and place them under the lamp. They do not need to eat or drink on their first day but make sure they have found the food and water on the following day. By observation, you will be able to see if the temperature is correct; the chicks will spread right out away from the lamp if it is too hot (in which case, raise the lamp a little higher) and crowd together underneath if too cool (in which case, lower the lamp a little closer to the chicks).

Each week, you'll need to raise the lamp a little until, by week three, the temperature is no higher than 23°C (75°F). During summer, chicks should no longer need a heat source once they are five weeks old; in colder weather, you'll need to keep the heat lamp on for another week or so. By the age of 6–8 weeks, the chicks are ready to move on to growers' pellets or feed. Start feeding them the layers' mash or pellets when they are ready to lay, at about 21 weeks.

When your hatch is successfully over, be sure to remove all shell debris from the incubator and rinse out the interior, making certain that the incubator is totally dry before closing it up.

Day-old chicks can successfully be introduced to a broody who has been sitting on sham eggs. Do this at night in as quiet a way as possible. Gently remove a couple of the sham eggs and pop a couple of day-olds in their place. Wait to see if the hen will accept them before continuing to remove the eggs and replace them with chicks.

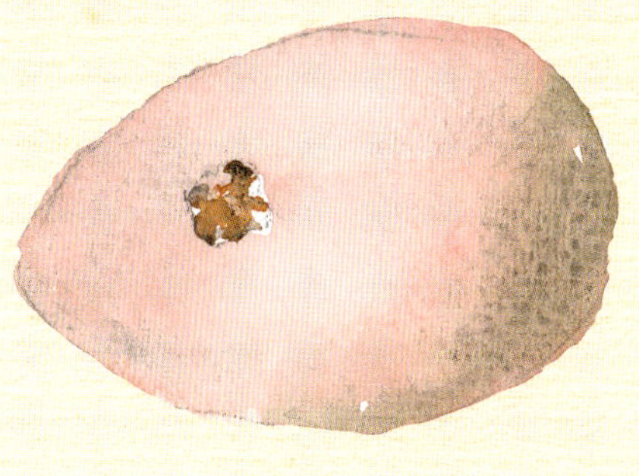

1

The egg tooth makes an initial hole.

2

The chick gradually works its way around the shell.

3

The shell begins to break apart.

4

The chick emerges from the shell.

5

Almost immediately, the chick stands up and its feathers will begin to dry.

# Parts of a chicken

**As you'll see** in the breed profile section (pp. 47–111), chickens vary widely in size and plumage. Yet there are common characteristics, as shown on page 38 and page 42. There are some notable differences between hens and cockerels.

- Saddle hackles and tail feathers tend to be pointier and longer in cockerels than in hens.
- Cockerels usually have thicker, sturdier legs.
- Wattles and combs tend to be brighter and larger in cockerels than in hens.
- Cockerels have spurs.

Dubbed game cock

# Combs

The comb is the red fleshy growth on the top of a chicken's head, and is usually larger in the cockerel. It is one of the distinguishing features of each breed, and comes in many varieties. Chickens cannot sweat, so blood pumping through the exposed comb and wattles naturally cools the rest of their body. The comb is also a useful signal of the overall health of the bird, and in the hen the comb indicates whether or not she is in lay. When in full lay, the comb will be large and bright red. When out of lay, the comb will be a paler pink and smaller in size than normal.

## Dubbing

Old English, Modern and certain other game bird breeds can only be shown if the cockerels have been 'dubbed'. Dubbing is the removal of the comb and wattles when the birds are approximately six months old. Combs are trimmed close to the line of the head and the wattles similarly to the throat, although each breed has its particular style and breed standards should be consulted. When the birds were used for fighting, dubbing was originally done to prevent injury, but nowadays it is purely cosmetic. In cold climates, it does help to protect them from frostbite. While still legal in the UK and US, this practice has now been banned in some European countries due to animal welfare concerns. If you decide to dub your birds, the procedure must be done under anaesthetic by a veterinarian.

## COMB TYPES

**Single:** the most common type of comb. Single combs come in many sizes, may be upright or flop to one or both sides and are usually larger in the cockerel. There may be different numbers of serrations or spikes.

**Rose:** the second most common type of comb. Rose combs cover the top of the head like a flat cap with a tapering spike at the back and are covered in small round knobbles. The spike may follow the line of the neck, be horizontal or turn up at the end.

**Pea or triple:** a low comb with three ridges, the middle one slightly higher than the other two and covered with small pea-like protuberances.

**V-shaped:** this is made up of two horn-like growths joined at the base.

**Walnut or cushion:** a small comb with no spikes or protuberances – it sits rather forward on the chicken's head.

**Cup:** this is really two single combs joined at the front and back and resembles a crown.

# Parts of a wing

# Feathers

Chicken feathers come in an astonishing range of colours and patterns, which help to make each breed recognisable. Plumage plays an important role, protecting the chicken from rain, cold and sun, and they must spend a considerable part of their time maintaining it. They do this by preening. Each feather has an axis or shaft onto either side of which the vanes are fixed; each vane has barbs on either side, which cling together but need to be 'combed' by the chicken, who also applies oil from a gland at the base of its tail. Dust bathing – which is when your chickens move or roll around in dust or dry earth to clean their feathers and skin – also plays an important part in maintaining their feathers.

Cockerels can be distinguished from hens by the fact that some of their feathers take on a different shape. Their hackle and saddle feathers are thinner and longer than a hen's; they also develop sickles, which are the spectacular curved feathers on either side of the tail – see Feather types below. Some breeds have much fluffier feathers than others, and game breeds have very tight feathering that often leaves a strip of bare skin down the breast. There may be feathering on the legs, and some breeds sport beards, muffs and crests.

Every year chickens moult – typically at the beginning of autumn – and replace their old feathers with new. As feathers are largely made up of protein, this takes a good deal of the hen's energy, so at this time it's important to give her plenty of replacement protein in the form of good-quality layers' ration. She will stop laying until her moult is complete, which could take anywhere between six and twelve weeks, and if the days are growing shorter, she may not start laying again until the days start to lengthen again after the winter solstice.

## FEATHER TYPES

## FEATHER MARKINGS

**Barring:** two distinct colours in bars across the feather – they may be regular or irregular and the width can vary.

**Double lacing:** as lacing, but with a second loop inside.

**Lacing:** a border of a different colour right around the edge of the feather – it may be broad or narrow.

**Frizzled:** each feather is curled, causing the bird to look distinctly unkempt.

**Mottled:** spotted in a different colour in a random fashion.

**Spangling:** a distinct contrasting colour at the end of the feather.

**Splash:** often drop-shaped marks of a contrasting colour in a random fashion.

**Pencilling:** how this looks depends on the breed – pencilling can look like a kind of barring or it can be like fine lacing.

**Peppered:** feathers have specks in a darker colour that look as though someone has ground pepper on them.

## FEATHER PATTERNS

**Birchen:** hackle, back saddle and shoulders white; neck hackles narrow black striping; breast black with silver lacing.
**Black:** male and female uniformly black with green sheen.
**Black mottled:** male and female black ground with white V-shaped tips on random feathers.
**Black red:** red hackles and black body and tail.
**Blue:** male and female uniformly slaty blue, head and neck may be darker; lacing, if present, darker.
**Buff:** male and female uniformly buff.
**Chamois:** male and female uniformly buff with paler lacing.
**Columbian:** male and female body mainly white; neck and tail black with some white lacing.
**Crele:** male hackles, back and saddle barred orange on pale ground; body barred grey and white. Female hackles barred greyish brown on pale ground; breast salmon; body as male.
**Cuckoo:** male and female dark grey to black indistinct barring on white ground; female can be darker than male.
**Exchequer:** male and female black and white randomly over body in blobs.
**Gold barred:** golden ground with distinct black barring.
**Gold spangled:** male and female hackle golden red with dark vane; body gold ground with black spangles; tail black.
**Jubilee:** male head, neck, body, legs and tail white; back and wings white with dark red markings. Female head and neck white; rest of body dark red with single or double lacing.
**Lavender:** male and female uniform slaty grey throughout.
**Mahogany:** male and female rich mahogany brown throughout.
**Millefleur:** male and female orange ground with black spangles with white highlights.
**Partridge:** male hackle, back and saddle greenish black with red lacing; breast and body black; female reddish lacing on black ground.
**Pile:** male head golden, hackle and saddle lighter; back red; front of neck white; wings mainly white. Female hackle white with gold lacing; neck and body white with salmon breast.
**Porcelain:** similar to Millefleur but bright beige ground.
**Quail:** complicated colouring giving impression that upper parts are dark and lower light; gold lacing and shafts.
**Red:** male and female bright red throughout.
**Silver barred:** male and female white to pale grey ground with bright black barring.
**Silver cuckoo:** male and female white to pale grey ground with dark grey to black indistinct broad barring.
**Silver duckwing:** male silver hackles and back; breast and body black; tail black with silver edging. Female silvery grey with salmon breast; tail and wings black with grey edging.
**Silver spangled:** male and female grey ground with black spangles.
**Speckled:** in speckled Sussex, male and female mahogany ground with white tips and black/green intermediate stripe.
**Splash:** male and female white ground with irregular slaty blue blobs, grey in places.
**Wheaten:** male gold hackles, rich brown body and dark green tail; female shades of wheat from golden to chestnut with black tips.
**White:** male and female uniformly white throughout.

# Breed profiles

Various breeds of pure-bred hens as well as hybrids are described with illustrations on the following pages. All individual breeds have different character strains within them, so please note that the descriptions given are intended to be used as a guide only, and where a hen is described as perhaps being flighty, this will be the general characteristic of that breed; certain lines of the same breed may be calm, but they will be in the minority.

The illustrations are typical types of each breed; however, if you are thinking of showing your birds then you should consult your country's breed standard.

For easy reference, the key details to consider for each breed are presented in a concise list.

**Purpose:** confirms a breed's primary purpose, i.e. whether a breed is suitable for laying eggs – 'layer' – or will lay few eggs but produce a fine carcass – 'table'. Some breeds may lay plenty of eggs and produce a good carcass – these are known as 'dual-purpose'. Breeds that are primarily bred for their appearance are called 'ornamental' or exhibition breeds, and these may not lay well or produce a good carcass.

**Class:** describes a breed's plumage and size. Game birds and Asian gamefowl are classed as 'hard feather' breeds, which means they have short feathers so tight to their bodies that, in some cases, the skin shows through. 'Soft feather' breeds have plumage that is looser and fluffier, and these breeds come in two sizes: 'light' and 'heavy', light being mainly birds of Mediterranean origin, which are excellent layers, and heavy being the larger breeds, which are frequently dual-purpose. Most large breeds have a bantam equivalent – 'true bantams' are birds that have no large equivalent.

**Broody:** indicates whether the breed tends to go broody frequently, occasionally or not at all.

**Number of eggs per year:** an estimate of how many eggs you can expect each breed to lay each year. Note that the actual number of eggs a hen will lay depends on many things, including food, comfort, their age, and so on. The terms included in the tables equate to the following ranges:

| | |
|---|---|
| Very few | 0–50 eggs per year |
| Few | 50–100 eggs per year |
| Moderate | 100–150 eggs per year |
| Prolific | 150–200 eggs per year |
| Very prolific | 200 or more eggs per year |

**Comb type:** indicates whether the breed has one or more of the six different types of comb – see Combs on pages 39–41.

**Feather colouring:** confirms when a breed comes in several different colourways – see also Feather markings on pages 44–45 and Feather patterns opposite.

Egg colour
The colour or colours of egg produced by the breed.
White
Off-white
Pale cream
Tinted
Pale tinted pink
Yellow
Brownish-yellow
Pinkish-plum
Light brown
Medium brown
Dark brown
Reddish-brown
Dark brown speckled
Blue
Khaki
Andalusians produce white eggs
Jersey Giants produce brown eggs
Wyandottes produce tinted eggs

# Ancona

Eye-catching, reliable layer that is happy in all but very confined situations

**PURPOSE:** Layer
**CLASS:** Soft feather light
**BROODY:** No
**NO. EGGS PER YEAR:** Prolific
**COMB TYPE:** Single
**EGG COLOUR:** White
**FEATHER COLOURING:** Black mottled

The Ancona is a tough, hardy bird that originated in the Italian port of its name, probably from Leghorn stock, in the mid-nineteenth century. It has beautiful black mottled feathers that have a beetle-green sheen in the sun and end with white tips. At first the white tips may not be very prevalent, but with each moult they become more so; this gives the birds a slightly camouflaged appearance, making them suitable for free ranging where predators may be a problem.

White earlobes denote a white-egg layer, and the Ancona produces a prolific number, even laying well through the winter. It tends, like other Mediterranean breeds, not to go broody. The single comb may droop to one side after the first point on the hen, though it is upright on the cock – rose comb varieties are also found.

This is an active bird, some might say flighty, that is happiest with plenty of space; any fencing will have to be high as they are also excellent flyers.

# Andalusian

An attractive bird that needs specialist breeding and is more suited to free range

**PURPOSE:** Layer
**CLASS:** Soft feather light
**BROODY:** No
**NO. EGGS PER YEAR:** Prolific
**COMB TYPE:** Single
**EGG COLOUR:** White
**FEATHER COLOURING:** Black, blue splash

Known as the Blue Andalusian in the United States, the Andalusian was developed in Spain in the mid-nineteenth century. Its beautiful blue plumage with slate-black lacing is produced by crossing a splash (blue-and-white mottled) cock with a black hen and doesn't breed true – i.e. two blue birds mated together produce a mixture of colours – although it is only the blues that are eligible to be shown. The bright red comb flops to one side after the first point and can suffer frostbite in cold climates – the bird being more suited to heat. It is an active forager that can run fast and tends to avoid human contact, which makes it difficult to tame.

This is not a beginner's bird, not only because of its flighty character but also because even if two blue birds are mated together the chances of breeding true blue offspring are small and most of the young stock will have black or white splashes in their plumage.

# Appenzeller

An exotic-looking bird that prefers plenty of space

**PURPOSE:** Ornamental layer
**CLASS:** Soft feather light
**BROODY:** Occasionally
**NO. EGGS PER YEAR:** Moderate
**COMB TYPE:** V-shaped
**EGG COLOUR:** White
**FEATHER COLOURING:** Many varieties including black, black mottled, chamois, gold spangled and silver spangled

The national breed of Switzerland, named after the strange-shaped bonnets worn by the local girls from the Appenzell area, which the bird's crest resembles. This is a hardy bird, active, and an accomplished flyer that likes its space and doesn't do well in confinement. If left to itself, it will happily roost up trees and seems able to survive snowstorms. It is a bird well adapted to mountain living as it has a very small comb and wattle, which reduce the risk of frostbite, and has tight feathering to retain heat.

Also known as the Appenzeller Spitzhauben, which translates as bonnet from Appenzell. There is another type, called Appenzeller Barthuhner or Bearded Hen, which is a larger version with a rose comb and beard but no crest. This breed was developed from crossing with Leghorns, Russian Bearded and a now-extinct type of Poland.

The Barthuhner lays a tinted rather than white egg but is similar in character to its more popular cousin.

# Araucana

A conversation-piece breed that lays exotically coloured eggs

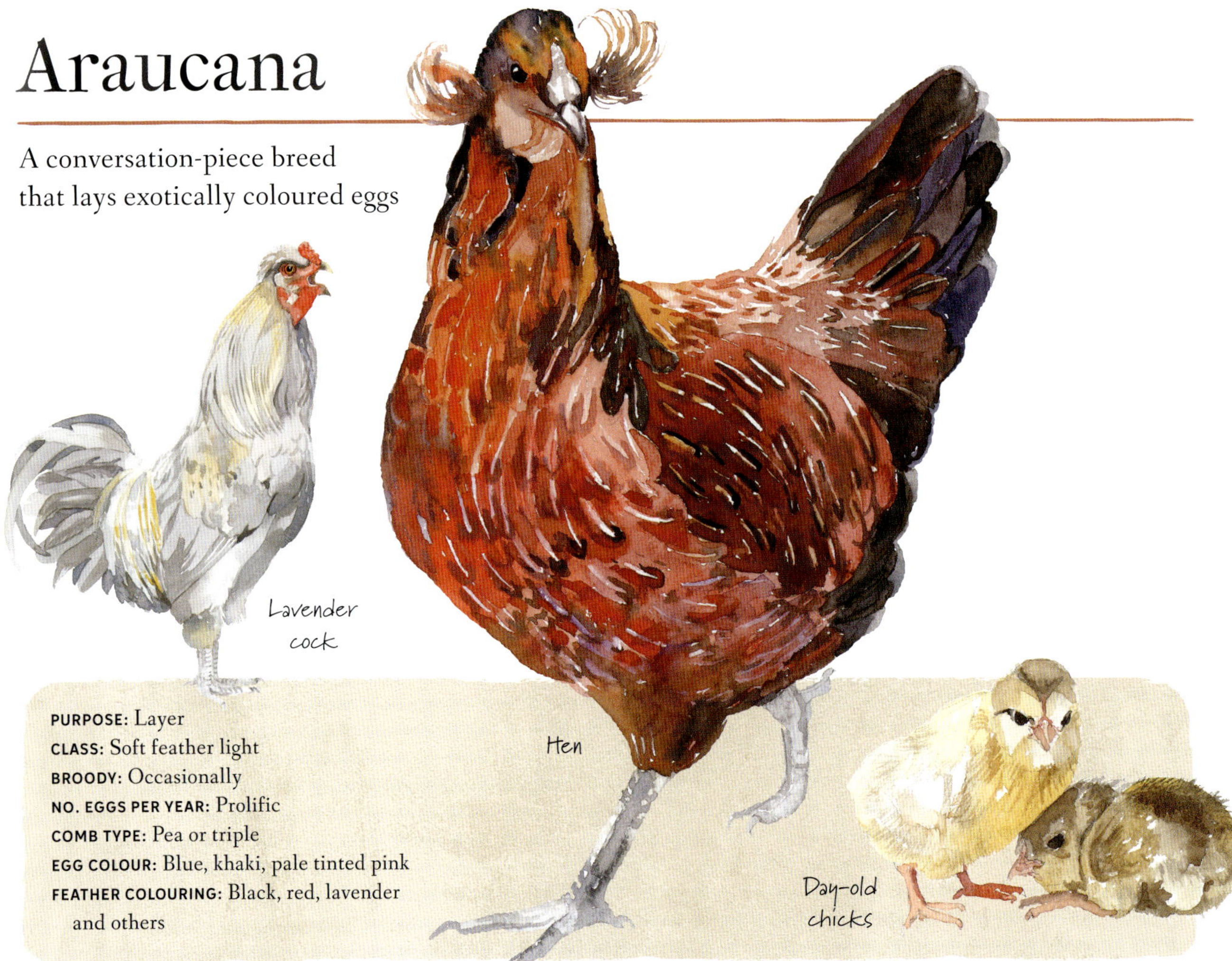


**PURPOSE:** Layer
**CLASS:** Soft feather light
**BROODY:** Occasionally
**NO. EGGS PER YEAR:** Prolific
**COMB TYPE:** Pea or triple
**EGG COLOUR:** Blue, khaki, pale tinted pink
**FEATHER COLOURING:** Black, red, lavender and others

This strange hen was named after the Araucania Indians of central Chile and can still be found in the wild in the Amazon basin. There is some dispute over how it arrived in the UK, but tales of blue-egg layers surviving shipwrecks in the Hebrides abound, and Araucanas are still popular in the Scottish Islands.

Araucanas have a rather upright stance with ear-tufts and crest and virtually no wattles; they are adaptable to confinement, do not show aggression and do well free range. Their eggs can range in colour from blue through greenish khaki to pink; hence the nickname 'the Easter egg chicken'. The birds come in many fairly random colours, but two lavenders mated together will produce lavender offspring.

In the United States, Araucanas are always rumpless, i.e. they have no tail; in fact the entire coccyx is missing and there is no uropygium or parson's nose, although this does not affect their laying ability. They also have pronounced ear-tufts, which uniquely grow from a lump of fleshy skin behind the earlobe; ideally these should point backwards.

The tailed version is known as an Ameraucana in the United States, and this should sport a muff and beard, along with a pea comb.

# Ardenner

A hardy bird, best suited to living free range

**PURPOSE:** Layer
**CLASS:** Soft feather light
**BROODY:** Occasionally
**NO. EGGS PER YEAR:** Prolific
**COMB TYPE:** Single
**EGG COLOUR:** White
**FEATHER COLOURING:** Many varieties

This old Belgian breed is a tough bird that has a reputation for laying well through the winter. Its wattles, face and single comb are a strange purplish colour, sometimes compared to the colour of mulberries or even blackberries. Ardenners come in a wide variety of colours such as partridge, white, black and golden-necked black, and silver and golden salmon. The birds' legs are also very dark, ranging from dusky to almost completely black. There is also a rumpless version of this breed (meaning the bird has no tail), which supposedly makes it harder for foxes to catch if kept free range, as there is nothing for them to get hold of. Indeed, these birds hate to be shut in and will roost in trees if given the chance. If left in the wild they would survive happily by themselves.

# Australorp

An excellent all-round bird for any situation

**PURPOSE:** Layer, table
**CLASS:** Soft feather light
**BROODY:** Yes
**NO. EGGS PER YEAR:** Very prolific
**COMB TYPE:** Single
**EGG COLOUR:** Light brown
**FEATHER COLOURING:** Black, blue laced and white

The Australorp was developed in Australia from black Orpington stock imported from England. One of this breed was reputedly a record layer in the 1920s, producing 364 eggs in 365 days.

The birds have a reputation for maturing early and laying on through the winter when other breeds stop. A large bird with bulging black eyes and somewhat fluffy feathering, they are hardy, docile creatures that are easily handled. They make very good mothers and can hatch up to 15 eggs at a time. The chicks, when born, have white down on their undersides, which gradually turns black as they develop.

# Barbu d’Anvers

A small, striking bird well suited to a small run

**PURPOSE:** Ornamental
**CLASS:** True bantam
**BROODY:** Yes
**NO. EGGS PER YEAR:** Moderate
**COMB TYPE:** Rose
**EGG COLOUR:** White
**FEATHER COLOURING:** Many varieties

This true bantam has existed in the Netherlands and Belgium since the seventeenth century, appearing in Dutch masters’ paintings of that time. Anvers is the French name for Antwerp and it can sometimes be called the Antwerp Bearded Bantam. Its beard, thick muff covering the earlobes and upright stance give it a striking appearance, causing it to constantly look as if it is about to crow. These bantams differ from the other Barbus in that they are always clean-legged.

Cocks can be aggressive in the breeding season but generally lack spurs. A special drinker will be required to keep the beard and muff dry.

Barbu du Grubbe is the name of the rumpless version created by a breeder of d’Anvers and named after their place of origin, Grubbe.

# Barbu d'Uccle

Pretty and popular breed that would grace any garden

**PURPOSE:** Ornamental
**CLASS:** True bantam
**BROODY:** Yes
**NO. EGGS PER YEAR:** Moderate
**COMB TYPE:** Single
**EGG COLOUR:** White
**FEATHER COLOURING:** Many varieties

Created by crossing a Booted Bantam with a Barbu d'Anvers in 1880, the Barbu d'Uccle has well-feathered legs. It differs from the Booted Bantam in that it also has a thick beard and muff and very small wattles, whereas the Booted Bantam has no beard and large wattles. These birds are good flyers and foragers but will need specialist care to keep the feathers in order – shelter from rain and no mud.

The fact that they have heavily feathered legs and feet makes this a suitable breed for keeping in a garden, as the feathering limits the amount of scratching they can do. This breed is sometimes known as Millefleurs (thousand flowers) or Millies as this is the commonest colouring, consisting of an orange/red ground with black-and-white mottled feathers.

# Barbu de Watermael

Small and lively but with a very loud crow

**PURPOSE:** Ornamental
**CLASS:** True bantam
**BROODY:** Yes
**NO. EGGS PER YEAR:** Few
**COMB TYPE:** Rose with three small spikes
**EGG COLOUR:** Pale cream
**FEATHER COLOURING:** Many varieties

Named after the Brussels suburb Watermael-Bosvoorde, this true bantam is similar to the d'Anvers, though lighter in frame. Being small, they are happy in very confined spaces and easy to tame but have an exceedingly shrill crow, beginning early in the morning, so a well-insulated henhouse is required unless neighbours are far enough away not to be disturbed.

They go broody frequently and make good reliable mothers, but are so small that they can't sit on more than seven of their own eggs and a maximum of five full-size eggs. They will require a water-tower drinker to protect their beard and muff.

# Barnevelder

A reliable, handsome layer of gorgeous dark eggs

**PURPOSE:** Layer
**CLASS:** Soft feather light
**BROODY:** Occasionally
**NO. EGGS PER YEAR:** Prolific
**COMB TYPE:** Single
**EGG COLOUR:** Reddish brown
**FEATHER COLOURING:** Black, double laced, partridge, white and silver

These fine layers were originally developed in the Netherlands to lay dark brown eggs. Many breeds were used in their development, including Marans to improve egg colour, which is a dark reddish brown, although the eggs tend to become lighter as the season progresses.

They have been found in the Barneveld municipality of the Netherlands since the twelfth or thirteenth century. Locals there claimed that their hens laid 313 eggs a year (being religious they rested on the Sabbath, or it would have been 365!).

A large, docile and friendly bird, Barnevelders are poor flyers and can be contained by even a low fence. They are known for their rather lazy disposition so would be ideal if only a small space was available – however, they can run to fat, which will affect their laying performance. They are hardy and are known to endure damp weather, continuing to lay through the winter, though it takes sunlight to show off the beauty of their greenish black lacing on bronze feathers.

Other colours found are black, partridge and white or silver, but it is the laced variety that is most popular and striking.

# Booted Bantam

An ideal pet particularly suited to life in a garden

**PURPOSE:** Ornamental
**CLASS:** True bantam
**BROODY:** Yes
**NO. EGGS PER YEAR:** Few
**COMB TYPE:** Single
**EGG COLOUR:** White
**FEATHER COLOURING:** Many varieties

Imported into the Netherlands in the sixteenth century from Java, this is a very old breed. They are sometimes called Sabelpoots in the Netherlands, Sabelpoot Kriel translating as 'sword-legged bantam', which refers to the very large 'vulture' hocks. These and the wide leg feathers need to be kept dry, so a mud-free covered area will be required. The heavily feathered feet supposedly discourage scratching, therefore limiting the damage that can be done to lawns and flower beds.

These very small but decorative bantams with delightful characters are popular in Germany and the Netherlands. They come in a wide variety of colours, some quite striking. Although they will spend a lot of their time broody, they are so small that even a child will be able to handle them, and make excellent pets if the number of eggs produced is not important.

# Brahma

The gentle giant of the hen world with a charming nature

**PURPOSE:** Layer, table
**CLASS:** Soft feather light
**BROODY:** Yes
**NO. EGGS PER YEAR:** Moderate
**COMB TYPE:** Large pea or triple
**EGG COLOUR:** Light to dark brown
**FEATHER COLOURING:** Buff, light, dark and other

Named after the Brahmaputra river in central and southern Asia, though in fact most likely created in the United States from Chinese and Indian birds. They were originally called Brahmapootras, which later was shortened to Brahma. These birds caused quite a stir when introduced to Great Britain in the 1850s. A number were given to Queen Victoria by an American breeder and were a great favourite of Prince Albert. Also known as chittagongs, the original birds all had dark plumage; the other colourings being developed over time.

Sometimes dubbed 'the King of Chickens', Brahma is one of the largest breeds of chicken in the world. This is a good-natured and docile hen that makes an excellent pet as it is extremely easy to tame if handled gently. The birds do not fly – they run – so do not need to be roofed in. However, this does make them unsuitable for small children who can easily be knocked over. They are best kept as trios, as one cockerel can only manage to fertilise a couple of hens rather than the normal six or seven. Broodiness occurs, but as the hens tend to get fat they often break the eggs

in their care, particularly as they lay surprisingly small eggs for such a large hen. They are slow to mature and may not lay their first egg until they are six or seven months old, but will continue to lay through the winter.

Blue partridge hen

An advantage of this breed is that not only are they tolerant of new introductions but the cocks also tolerate each other and are not ardent crowers.

# Cochin

Attractive additions to any garden, with many uses

Black hen and cock

**PURPOSE:** Layer, table, ornamental
**CLASS:** Soft feather light
**BROODY:** Yes
**NO. EGGS PER YEAR:** Moderate
**COMB TYPE:** Single
**EGG COLOUR:** Light brown, yellow
**FEATHER COLOURING:** Black, blue, buff, partridge, white and others

Cochins were originally known as Shanghai Fowl when imported from China around 1840. Queen Victoria was presented with seven of these exotic birds. No one had seen anything like them, so different from the farmyard fowl native to Britain. A special aviary was built for the influx of exotic poultry which the Queen delighted in. She sent fertile eggs to her royal relatives. The breed eventually reached the United States in the 1870s.

One of the largest breeds – cocks can reach 5kg (11lbs) – it was the Cochin that launched interest in poultry shows, causing a sensation because of its huge size. This really is an all-round bird, as not only does it produce a fine carcass and a moderate number of eggs but also it has very soft feathers, which were used to stuff pillows and feather mattresses.

Being so large, these birds are slow to mature, but eventually grow into very fluffy, peaceful and friendly creatures, with feathered legs and feet and small, rounded, downy tails.

They need little space, and because of their feathery feet tend not to scratch as much as some;

however, they will need a mud-free sheltered area to retain the quality of their feathers. For a bird of such great size, they lay surprisingly small eggs. The hens make excellent mothers and frequently go broody.

Although Pekins (pp. 86–87) look like bantam Cochins, they are in fact recognised as a separate breed.

# Cream Legbar

An auto-sexing layer of beautiful blue eggs

**PURPOSE:** Layer
**CLASS:** Soft feather light
**BROODY:** Yes
**NO. EGGS PER YEAR:** Prolific
**COMB TYPE:** Single
**EGG COLOUR:** Blue and green
**FEATHER COLOURING:** Gold, silver and crele

The Cream Legbar is what is known as an auto-sexing breed, i.e. when the eggs hatch there is a marked difference between the male and female chicks (males being paler). This breed was developed from many different breeds, but includes Araucana, from where it gets its beautiful blue eggs and crested head. In front of the crest the single comb flops over, often in two directions. They are alert, inquisitive birds that tend not to make the most reliable of mothers.

Cream Legbar are excellent foragers. The downside of this is that they will scratch up large areas of your garden, should you have them free ranging. They are lively, flighty birds – an advantage where there are predators present as they will rapidly fly up into a tree or henhouse roof to escape.

The Legbar family also includes rarer breeds such as Cambars, Rhodebars, Welbars and Wybars, all of which have barred Plymouth Rock in their make-up and all of which are auto-sexing.

Hen
Hen

# Croad Langshan

A handsome, docile breed with exceptionally lustrous plumage

**PURPOSE:** Layer, table
**CLASS:** Soft feather heavy
**BROODY:** Occasionally
**NO. EGGS PER YEAR:** Prolific
**COMB TYPE:** Single
**EGG COLOUR:** Pinkish plum
**FEATHER COLOURING:** Black, white

Imported from China by Major F. T. Croad in 1872 and named after the Langshan district of the Yangtze (Chang) river, this bird resembles the Cochin, and at first there was some argument as to whether they were the same breed.

These birds are exceptionally tall, with lustrous black plumage that appears bottle green in the sun, and they have a single comb with five points. Their height is accentuated by their high tail carriage. Their legs should be feathered down the outside, and unusually their feet have pink soles – black spots in the pink skin is a defect. They are slow to mature, intelligent, adaptable and cold hardy.

This breed is not recognised in the United States, but a slightly different version including Minorca and Plymouth Rock blood, known simply as Langshan, is.

# Derbyshire Redcap

An ancient breed – one for the wide-open spaces

**PURPOSE:** Layer, table
**CLASS:** Soft feather light
**BROODY:** No
**NO. EGGS PER YEAR:** Prolific
**COMB TYPE:** Large rose
**EGG COLOUR:** White
**FEATHER COLOURING:** Red, black

Derbyshire Redcaps were developed, as their name suggests, in Derbyshire from a Hamburgh cross and are thought to be the oldest breed of domestic fowl in Great Britain. They are mostly known just as Redcaps, being named for their huge rose comb, which is much larger than that of any other breed.

They are active, characterful birds, quite capable of finding a greater part of their food outside the enclosure if allowed to forage, and are also excellent flyers. Derbyshire Redcap is not a breed that would be happy in a small enclosure, as it needs a large space to show off its lustrous mahogany feathers tipped with black spangles.

# Dominique

An attractive dual-purpose bird with smart barred plumage

**PURPOSE:** Layer, table
**CLASS:** Soft feather heavy
**BROODY:** Occasionally
**NO. EGGS PER YEAR:** Moderate
**COMB TYPE:** Rose
**EGG COLOUR:** Brown
**FEATHER COLOURING:** Black-and-white barred

The Dominique was originally the most popular breed in the United States, documented as far back as the mid-eighteenth century, where they were not only a dual-purpose breed but also their feathers were used for stuffing pillows and feather beds. They were instrumental in the development of the barred Plymouth Rock in the mid-nineteenth century, which superseded them in popularity.

Dominiques will tolerate confinement but can be flighty. They do best if allowed to range, as they are good hardy foragers, and their cuckoo-barred plumage, also known as 'hawk colouring', supposedly gives them a certain protection from airborne predators.

A useful attribute is that, with practice, the chicks' sexes can be told apart on hatching, as the cockerels have a small, scattered spot of yellow on their heads, and in the hens the spot is more apparent.

# Dorking

The archetypal English hen – did it come over with the Romans?

**PURPOSE:** Layer, table
**CLASS:** Soft feather heavy
**BROODY:** Occasionally
**NO. EGGS PER YEAR:** Prolific
**COMB TYPE:** Single in red and silver grey, rose in cuckoo and white, single or rose in dark
**EGG COLOUR:** White
**FEATHER COLOURING:** Cuckoo, dark, red, silver grey, white

Dorkings were thought to have been brought to England by the Romans; there is a description of a hen with five toes by a Roman farmer and agricultural historian, Columella. The five toes are an important breed characteristic, although the fifth toe appears to have no purpose and is found just above the fourth toe. Named after the town of that name in Surrey, Dorkings were originally bred as table birds with very white meat and supposedly had no equal. They reached the United States by travelling from England with early settlers.

This is a breed that doesn't like to be too confined and would be happiest scratching about in a farmyard – indeed, they like to range widely. Having said that, they have rather short legs and seem to scratch less than other breeds, therefore making them very suitable for a garden environment. Thanks to their rather short legs and fluffy plumage, they tend not to do well in damp or muddy conditions. They will get dirty and

*Continues overleaf*

Dorking, *continued*

bedraggled very quickly – not a good look if showing is your aim. They are rather slow to mature and may not start laying until at least 26 weeks old, but once they start they have a reputation for laying well through the winter. They only occasionally go broody but make good mothers, and being gentle souls are prone to bullying from more outgoing breeds.

These are large birds that will need plenty of space in the henhouse. You should ensure that nest boxes are large enough to fit a plump hen or her beautiful plumage will be damaged. Dorkings will also appreciate a lower roosting bar – easier for short legs and heavy bodies to reach.

# Faverolles

A gentle breed, easy to handle with a calm disposition

**PURPOSE:** Layer, table
**CLASS:** Soft feather heavy
**BROODY:** Occasionally
**NO. EGGS PER YEAR:** Prolific
**COMB TYPE:** Single
**EGG COLOUR:** Light brown
**FEATHER COLOURING:** Salmon, but others found

This is the heaviest of the French breeds, developed in the nineteenth century from a cross that included Houdans and Dorkings, from which the Faverolles got its distinctive five toes, muff and beard. This fast-growing breed is famed for the fine texture of its meat and also the fact that it continues to lay through the winter, when some breeds stop.

Faverolles make excellent pets, being calm, docile and gentle, but are therefore easily bullied by more energetic breeds. They seem quite content in fairly small runs. Various colourings are found, but salmon is the commonest and in fact the only colour recognised for showing by the APA (American Poultry Association).

# Hamburgh

A self-sufficient and smart breed – one for the open spaces

**PURPOSE:** Layer
**CLASS:** Soft feather light
**BROODY:** Occasionally
**NO. EGGS PER YEAR:** Prolific
**COMB TYPE:** Rose
**EGG COLOUR:** White
**FEATHER COLOURING:** Black, gold pencilled, gold spangled, silver pencilled, silver spangled, white

The origins of the Hamburghs (or Hamburg in the United States) have become lost in the mists of time. As they were also called Mooneys or Hollands, some thought they were developed by the Dutch, but it is more likely that they originated in Eastern Europe. Regardless, they have been known in Great Britain for at least 300 years. They are small, active, graceful birds with a smart rose comb and upright tail.

Hamburghs were originally bred as fighting birds and still have a very aggressive streak. Indeed, if they are too enclosed and bored they will even attack each other. This is not a friendly breed and prefers its own company to that of humans. They are unreliable mothers but do lay a good quantity of small white eggs. They are beautiful birds, whether spectacularly pencilled or spangled, and their strong flying ability can be curbed by the clipping of one wing.

# Ixworth

An unusual dual-purpose bird restricted to the UK

**PURPOSE:** Layer, table
**CLASS:** Soft feather heavy
**BROODY:** Yes
**NO. EGGS PER YEAR:** Moderate
**COMB TYPE:** Pea or triple
**EGG COLOUR:** Tinted
**FEATHER COLOURING:** White

Reginald Appleyard, well known for the Silver Appleyard duck, is credited with producing the Ixworth in the Suffolk village where he lived in 1932.

What he tried to create was a table bird with white skin that also laid well and matured fast, rather similar to the Canadian Chantecler. Breeds included in its make-up were white Sussex, white Orpington and several colours of Indian Game. Sadly the Ixworth was never exceptional at laying or as a table bird, although it does have a devoted following in the UK, particularly with organic/free-range producers.

The Ixworth is a handsome well-built bird. It possesses white skin, typically favoured over yellow-skinned breeds. It is only found with white plumage but this is set off with a bright pink pea comb and beak, and exceptionally pink legs.

# Japanese Bantam

Ornamental bird suited to small enclosures and junior handlers

Frizzled cock

Hen

Day-old chick

**PURPOSE:** Ornamental
**CLASS:** True bantam
**BROODY:** Yes
**NO. EGGS PER YEAR:** Few
**COMB TYPE:** Single
**EGG COLOUR:** Brown, cream or white
**FEATHER COLOURING:** Many varieties

The Japanese Bantam, or Chabo as it is also known, is a true bantam, i.e. it has no large equivalent. It is acknowledged to have existed in Japan in the 1650s, as it appears in art of that time, and reached the West shortly after, as it also appears in seventeenth-century Dutch paintings. These bantams were kept as ornamental garden birds by Japanese society, who bred them as a hobby.

There are various types of this breed, some of which typically have exceptionally large combs, some have frizzled feathers and all have legs so short they are almost invisible, with wings that hang down to touch the ground, giving the appearance that the creature is sitting. The characteristic tail stands up over the body. In order to keep the fancy feathering clean, they should only go out in dry weather; in any case they are not cold hardy and should be kept inside in winter as their breasts brush the ground and they get easily chilled if wet.

Exceptionally gentle and submissive, they make charming pets and are very suitable for children to handle, their short legs making them rather slow moving. They will live happily in a small enclosure but lay only a few very small eggs.

# Jersey Giant

A loveable American giant with a loveable character to match

**PURPOSE:** Layer, table
**CLASS:** Soft feather heavy
**BROODY:** Yes
**NO. EGGS PER YEAR:** Prolific
**COMB TYPE:** Single
**EGG COLOUR:** Brown
**FEATHER COLOURING:** Black, blue laced, white

As its name suggests, this breed was developed in New Jersey in 1870, by crossing Brahmas with Langshans and the Java. Although very slow to mature, Jerseys were very popular as capons (birds that are neutered to improve the quality of the flesh for eating), before caponising was banned, some reaching an incredible 9kg (20lbs) or so. If allowed to grow on naturally, a good cock bird will reach a respectable 6kg (13lbs).

Jersey Giants make excellent mothers, being calm, gentle birds. Owing to their great size they are almost unable to fly, so do not need high fencing to contain them, but they do need plenty of space and a large henhouse to accommodate them. If breeding for the table, the earlier in the year hatching is achieved the larger the birds will be. The brown eggs are also exceptionally large and the hens tend to have a long laying season.

Black was the colour first developed and is still the most popular, having a wonderful greeny-coppery sheen in the sun, but they also come in white, and – rarely – blue laced.

# Leghorn

An excellent, reliable layer in a colour of your choice

**PURPOSE:** Layer
**CLASS:** Soft feather light
**BROODY:** No
**NO. EGGS PER YEAR:** Very prolific
**COMB TYPE:** Single, rose
**EGG COLOUR:** White
**FEATHER COLOURING:** Many varieties

The Leghorn (pronounced 'legorn' with silent 'h') comes in a wider variety of colours and form than any other breed of poultry. It originated in the Italian city of Livorno (leghorn being the English translation of Livorno) and due to its remarkable laying ability is now found all over the world, where it has been developed into a number of different strains, all stemming from the same initial stock.

All birds have white earlobes and yellow legs in common, mature early, and lay their large white eggs through the winter. They will readily scavenge for themselves if left free to roam but also take happily to life in a run. Cockerels have an exceptionally large upright comb which falls elegantly to one side in the hen. As with most Mediterranean breeds, they go broody only very rarely.

Leghorns tend to be a flighty breed and the birds are good fliers. The upside of this is that they are capable of protecting themselves against predators. The downside is that they may prefer to roost in trees, which means if you wish to keep them confined, wing clipping will be required. Leghorns are hardy birds and can tolerate heat as well as cold.

Light brown hen

# Marans

An excellent backyard layer of spectacular dark brown eggs

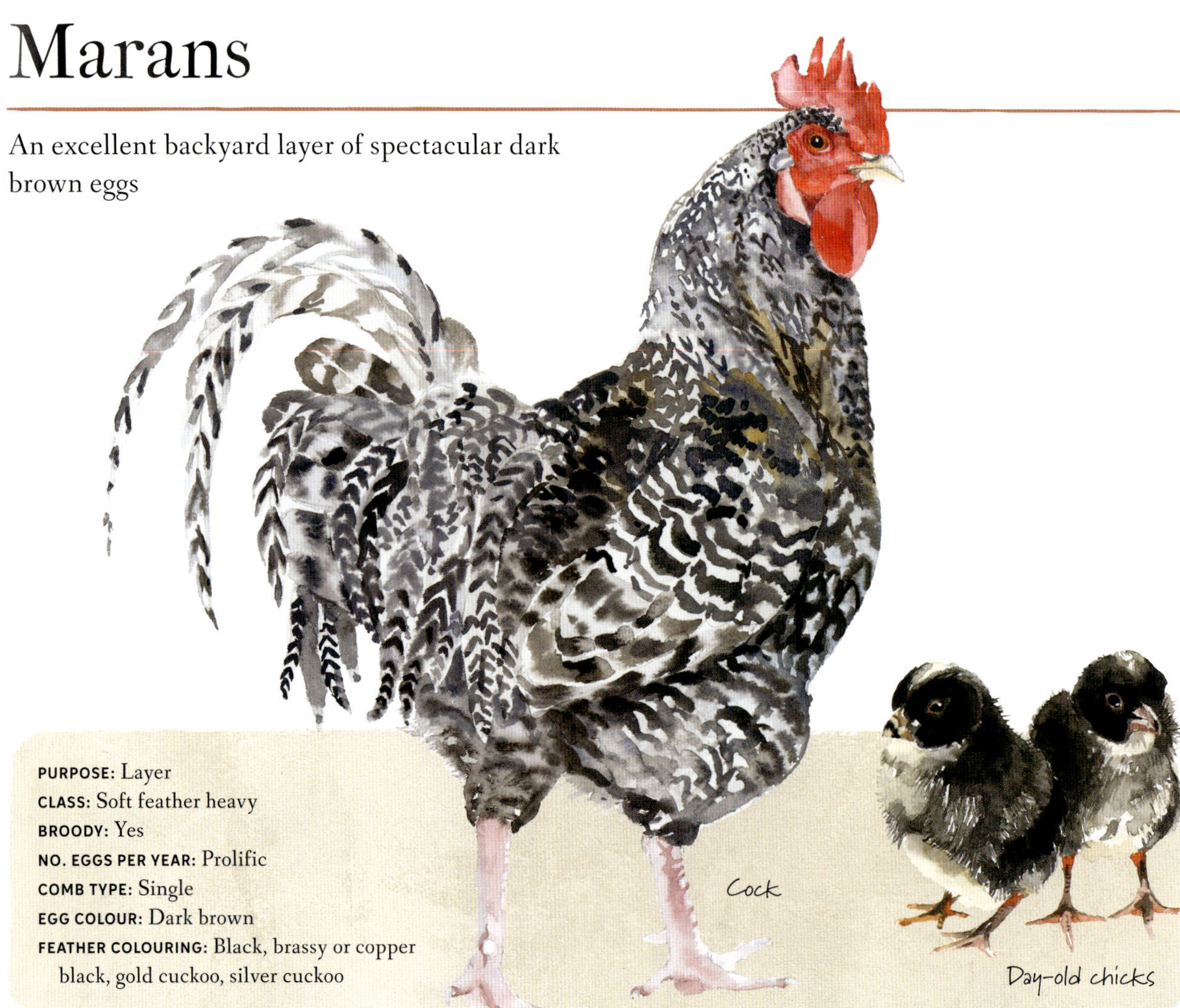


**PURPOSE:** Layer
**CLASS:** Soft feather heavy
**BROODY:** Yes
**NO. EGGS PER YEAR:** Prolific
**COMB TYPE:** Single
**EGG COLOUR:** Dark brown
**FEATHER COLOURING:** Black, brassy or copper black, gold cuckoo, silver cuckoo

The Marans was developed in the northern French fishing village of its name in the early twentieth century from a crossing of several breeds, but in particular the Langshan. The original Marans were first introduced to Great Britain in the 1920s by Lord Greenway, a British businessman who imported fertile eggs after seeing them at a Paris show.

Originally the French birds had lightly feathered legs inherited from the Langshan, but this was bred out of the British and American lines as clean legs are preferred, being less susceptible to scaly leg mite. Marans lay remarkably dark, chocolate-brown eggs that have thick shells with very small pores; it is thought that this has the advantage of preventing salmonella bacteria entering the egg.

Although certain breed lines are inclined to be rather flighty, in the main they are friendly birds and easy to tame. They make excellent mothers and the chicks, when hatched, can be sexed by an experienced eye, as the cockerels tend to be lighter in colour than the

pullets. Along with other barred breeds, if the female is mated with a compatible non-barred cockerel, sex-linked chicks will result, the males having a larger white head-spot.

Marans have yet to be recognised by the APA but have a growing following in North America.

# Marsh Daisy

One to consider if your ground tends to be boggy

**PURPOSE:** Layer
**CLASS:** Soft feather light
**BROODY:** No
**NO. EGGS PER YEAR:** Moderate
**COMB TYPE:** Rose
**EGG COLOUR:** Tinted
**FEATHER COLOURING:** Black, buff, brown, wheaten, white

Cross an Old English Game cock and Malay hen, add a bit of Hamburgh and Leghorn and finally stir in some Sicilian Buttercup, mix well and you will have created a Marsh Daisy. This took place in Lancashire in the 1880s, and such was their popularity that a breed club was formed shortly after the First World War.

Sadly, the Marsh Daisy is now fairly rare, which is a pity as it is a beautiful, brightly coloured bird with a handsome rose comb, particularly large on the cockerel. It is also a good forager that has the reputation of thriving on swampy ground and in wet conditions, making it sound as if it is aptly named – in fact the 'Marsh' comes from Marshside, where it was developed. Unusually, there is no bantam version of this breed.

# Minorca

A handsome layer of large white eggs that was known in Roman times

**PURPOSE:** Layer
**CLASS:** Soft feather light
**BROODY:** No
**NO. EGGS PER YEAR:** Very prolific
**COMB TYPE:** Single
**EGG COLOUR:** White
**FEATHER COLOURING:** Black, buff, white

It is unknown if this particular breed did originate on the Balearic island of Menorca (also known as Minorca), but what is known is that some form of it was around in the time of the Romans. The present-day form was developed in England into the heaviest of the light breeds.

This is a bird that likes the heat; it will tolerate cold, but its exceptionally large comb, wattles and earlobes make it vulnerable to frostbite. The hen's comb droops gracefully to one side but should not obstruct her vision. A rather flighty creature that needs tactful handling, the Minorca will reward its owner with a good number of large white eggs.

# New Hampshire Red

Ideal all-round bird with attractive chestnut plumage

**PURPOSE:** Layer, table
**CLASS:** Soft feather heavy
**BROODY:** Yes
**NO. EGGS PER YEAR:** Prolific
**COMB TYPE:** Single
**EGG COLOUR:** Light to medium brown
**FEATHER COLOURING:** Reddish brown

The New Hampshire Red was developed in the early twentieth century by researchers and farmers in New Hampshire directly from Rhode Island Red stock by careful selection of early-maturing, vigorous layers of large brown eggs. What they produced was an adaptable, friendly bird that lays well and produces a good-size carcass. It was admitted to the APA in 1935.

This breed is popular in Germany and the Netherlands, and since it arrived in the UK in the early 1980s, it has attracted a great deal of attention.

The New Hampshire Red differs from the Rhode Island Red in its colouring, being a much lighter, more chestnut brown compared with the Rhode Island Red's mahogany colouring, and also has a different body shape with higher tail carriage.

# Norfolk Grey

An unusual choice for a dual-purpose bird

Hen

**PURPOSE:** Layer, table
**CLASS:** Soft feather heavy
**BROODY:** Yes
**NO. EGGS PER YEAR:** Moderate
**COMB TYPE:** Single
**EGG COLOUR:** Tinted
**FEATHER COLOURING:** Black

This breed was developed by Fred Myhill of Norwich around 1920. There is a similar-looking German breed called Birchen Niederrheiner, which is not related. Originally, they were known as Black Marias, but Mr Myhill thought the connotations too unpleasant and changed it.

Although fairly popular in the 1920s, they dwindled in numbers until re-emerging in the 1970s, and can now be found in plentiful numbers in Norfolk and occasionally in other parts of the country. They are an easy-to-keep, dual-purpose bird with attractive plumage.

# Orpington

A large cuddly ball of fluff with a docile and affectionate manner

**PURPOSE:** Layer, table
**CLASS:** Soft feather heavy
**BROODY:** Yes
**NO. EGGS PER YEAR:** Prolific
**COMB TYPE:** Single
**EGG COLOUR:** Pinkish brown
**FEATHER COLOURING:** Black, blue, buff, white

William Cook of Orpington, Kent, who was editor of the *Poultry Journal*, created this breed in 1880 with help from his daughter. The idea was to produce a prolific layer and in this the Cooks were successful as Orpingtons can lay as many as 340 eggs per year, a remarkable feat. What they made was a cold-hardy, adaptable, docile and affectionate creature that is easy to handle, but its compliant nature makes it all too open to being bullied by other more ebullient breeds.

A black Orpington pullet won the grand prize at the Crystal Palace Poultry Show in 1886; however, gradually over the years selection of birds for looks rather than egg production has resulted in declining laying records. A good heavy carcass is still produced, though the full feathering makes the birds look rather larger than they actually are. Nevertheless, their size must be considered and extra space in the henhouse will be required. Orpingtons are enthusiastic and reliable mothers and can cover many eggs at a sitting, including goose eggs. Due to their ample feathering, they may suffer in hot temperatures and will need plenty of shade in their run.

Orpingtons have a loyal following – a flock was owned by the late Queen Elizabeth, as well as the Queen Mother – and those who settle for this breed rarely seem to find a need to try others.

# Pekin

A large cuddly ball of fluff with a docile and affectionate manner

Lavender cock

**PURPOSE:** Ornamental
**CLASS:** True bantam
**BROODY:** Yes
**NO. EGGS PER YEAR:** Few
**COMB TYPE:** Single
**EGG COLOUR:** White or pale cream
**FEATHER COLOURING:** Many varieties

Day-old chick

The Pekin is what is classed as a true bantam, which means it has no large equivalent. It looks very like a miniature version of a Cochin but is classed as a different breed in the UK; they are known as Cochin bantams in the United States. They are said to have been imported by soldiers looting the Emperor of China's palace in Peking in 1860.

Pekins have very soft fluffy feathering, with the top of the tail being higher than the head and no legs or feet visible, although these are feathered. The fact that the feathers are dragging on the ground means that they need to be kept out of muddy areas but also that they won't scratch about quite as much as some other breeds and very rarely fly. They go broody frequently and make excellent mothers. One other attribute that makes them ideal as pets is that their crow is a little quieter than most.

Hen

# Plymouth Rock

The archetypal all-round hen – most popular of all

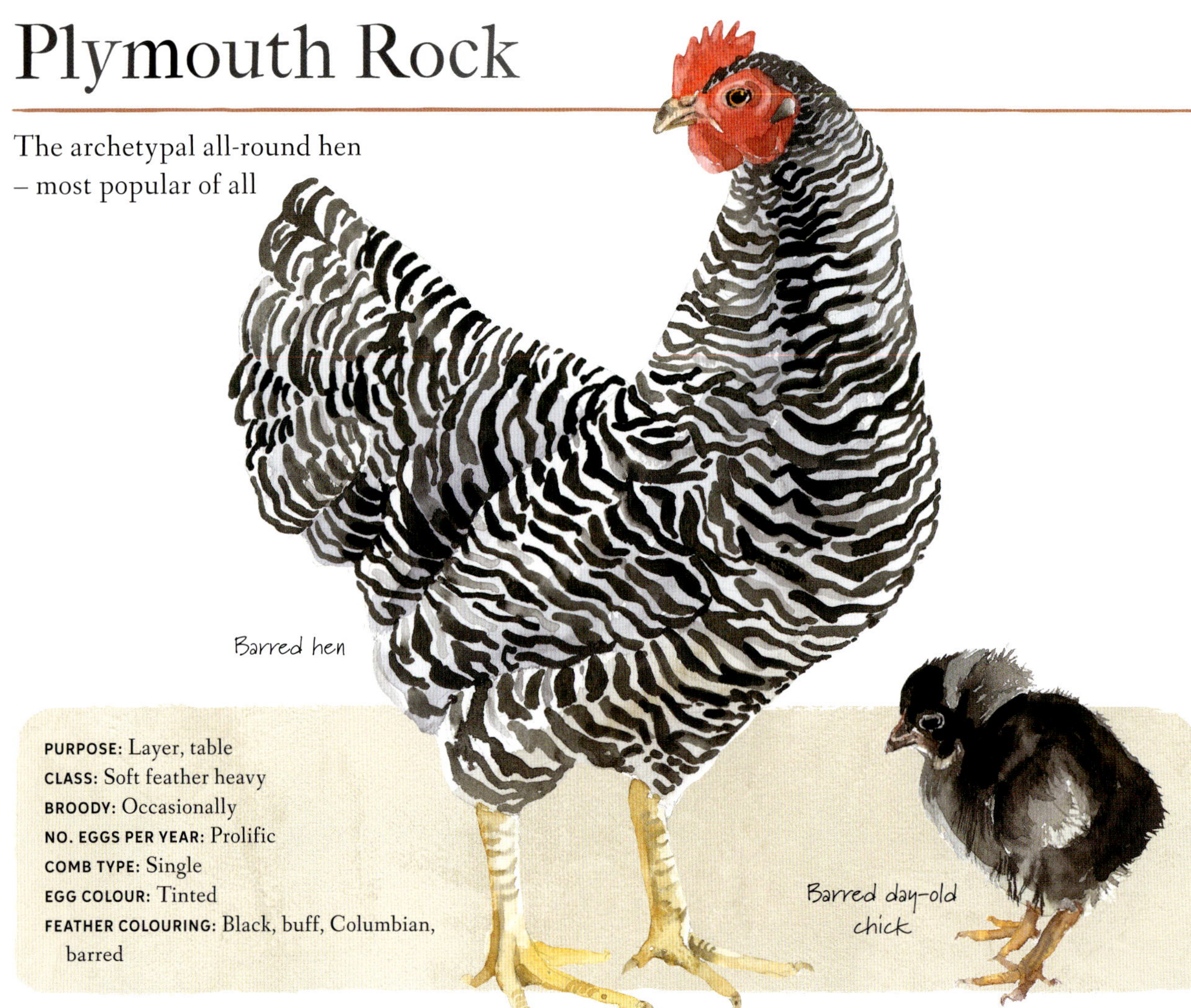


**PURPOSE:** Layer, table
**CLASS:** Soft feather heavy
**BROODY:** Occasionally
**NO. EGGS PER YEAR:** Prolific
**COMB TYPE:** Single
**EGG COLOUR:** Tinted
**FEATHER COLOURING:** Black, buff, Columbian, barred

Plymouth Rock is a granite boulder on the shore of Plymouth Bay, Massachusetts, where the Pilgrim Fathers stepped ashore in 1620, but it wasn't until the mid-nineteenth century that Plymouth Rocks as we know them today were created from, among others, crossings with Dominiques, Brahmas and Cochins.

They are now the most popular dual-purpose breed in the United States, and no wonder, as they are robust birds that mature early, are content to free range or be kept confined and are easily handled. Plymouth Rocks are long-lived, hardy in cold weather and unaffected by hot temperatures. They are heavy birds and only fly if absolutely necessary, so fencing need not be so high to keep them contained.

The cocks have the reputation of being even-tempered and less aggressive than some breeds. Although found in many colour variations, the barred version is the most commonly seen, and all colours have distinct yellow legs.

Barred cock

# Poland

A placid individual with a pompom on top

**PURPOSE:** Ornamental
**CLASS:** Soft feather light
**BROODY:** No
**NO. EGGS PER YEAR:** Very variable
**COMB TYPE:** V-shaped
**EGG COLOUR:** White
**FEATHER COLOURING:** Chamois, golden, silver, white, white-crested black

Chamois hen

Known as Polish in the United States, the Poland is undoubtedly a very old breed and can be seen in Dutch paintings dating from 1600. It seems very unlikely that it originated in Poland and the name was a corruption of something else – probably 'polled'. Another theory is that the name came from the fact that the chicken's crest feathers resemble the feathered helmets once worn by Polish soldiers.  Whatever its origin, this is a very ancient breed. There were Polands at the very first poultry show in London in 1845 and the breed was standardised in the first Book of Standards in 1865. Polish chickens were recognised by the APA in 1874.

Their rather comic appearance can have its disadvantages: the enormous crest limits the bird's vision, allowing it only to see forward and down; owners need to warn it of their approach or it is easily frightened. Being out of reach for preening, the crest is prone to lice.

The crest appears as the chickens have a vaulted skull (also known as a cerebral hernia or knobbed skull). This is a genetic mutation that causes a bulge or protrusion on the top of the skull.

In order to keep this fancy feathering in good condition, a certain amount of cover from rain will be required, but these birds are quite content in confinement, being friendly and gentle, but tend to be bullied by more energetic breeds.

# Rhode Island Red

Exemplary beginners' backyard bird – the best of both worlds

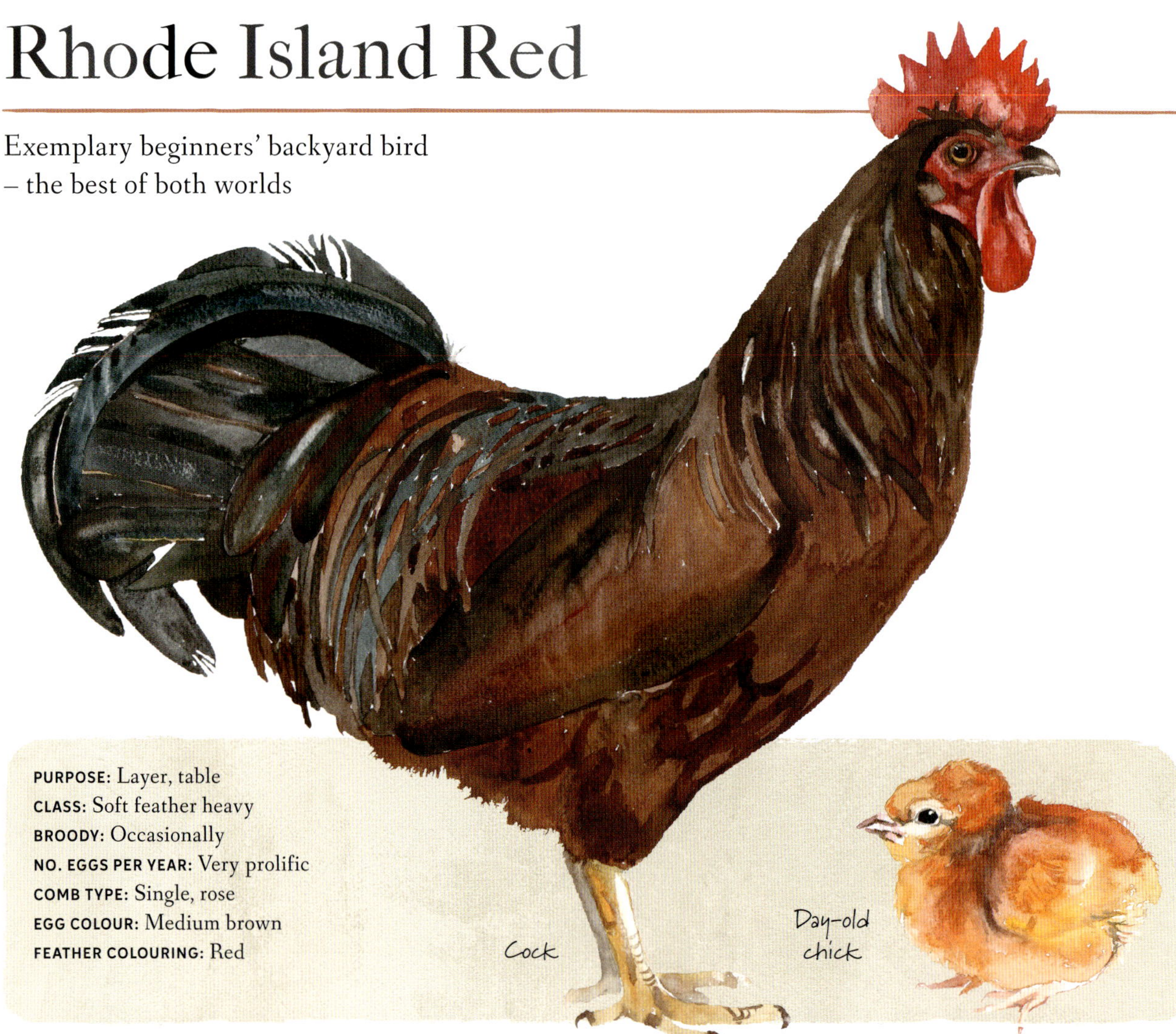


**PURPOSE:** Layer, table
**CLASS:** Soft feather heavy
**BROODY:** Occasionally
**NO. EGGS PER YEAR:** Very prolific
**COMB TYPE:** Single, rose
**EGG COLOUR:** Medium brown
**FEATHER COLOURING:** Red

The quintessential hen, state bird of Rhode Island and arguably the most successful breed ever created. What most people consider a traditional hen was created by crossing Asiatic fowls such as Shanghai, Malay and Java to create an excellent dual-purpose breed with a reputation for reliable egg-laying and a fine carcass, along with a certain resistance to disease.

As with many breeds, the cocks can be aggressive, but as a rule the hens are calm, easy to handle and will live happily in a large run or free range. They are not spectacular flyers so can be contained by a low fence.

Rhode Island Reds are often used as a parent breed in creating hybrids, particularly with Plymouth Rock. Known for their reliable egg production and hardiness, these hybrids are popular with poultry keepers.

Hen

# Rhode Island White

A handsome all-round bird with an easy-going character

**PURPOSE:** Layer, table
**CLASS:** Soft feather heavy
**BROODY:** Occasionally
**NO. EGGS PER YEAR:** Very prolific
**COMB TYPE:** Rose
**EGG COLOUR:** Medium brown
**FEATHER COLOURING:** White

This is not a white Rhode Island Red but a separate breed altogether, and does not enjoy the popularity of the former. It is made up of a mix of Wyandottes, Cochins and Leghorns and was created around 1900.

When crossed with a Rhode Island Red it produces the classic sex-link, i.e. the female chicks are red and the males white. It is a calm, hardy bird with an easy-going character, which matures early, is an excellent all-rounder and would make a brilliant choice as a slightly more unusual backyard bird.

# Rosecomb

King of the show ring, and handsome to boot

**PURPOSE:** Ornamental
**CLASS:** True bantam
**BROODY:** Occasionally
**NO. EGGS PER YEAR:** Very few
**COMB TYPE:** Rose
**EGG COLOUR:** White or pale cream
**FEATHER COLOURING:** Black or other variations

Rosecombs are true bantams, i.e. they have no large equivalent. Their origin is unclear but they have been found in the British Isles since the fifteenth century. They are smart little birds, with their rose combs that have a long leader or spike, and large white earlobes.

Laying is not their strong point – their eggs are exceptionally small – and they are bred mainly as pets. In the breeding season cocks can be aggressive, although the hens are friendly and easy to tame. They take kindly to confinement, although they love to forage and if free range it must be remembered that they are accomplished flyers. Thanks to a certain amount of inbreeding, they can suffer from infertility, which makes breeding difficult.

# Scots Dumpy

An endearing personality with a mothering instinct

**PURPOSE:** Layer, table
**CLASS:** Soft feather light
**BROODY:** Yes
**NO. EGGS PER YEAR:** Moderate
**COMB TYPE:** Single
**EGG COLOUR:** White
**FEATHER COLOURING:** Cuckoo most common, but many other varieties including black and white

Hen

The exceptional feature of this breed is its very short legs, giving rise to a long list of other names, such as creepers and crawlers. It is an ancient breed, and supposedly alerted the Scots and Picts to potential attacks from the Romans due to its acute hearing and alarm calls.

By the middle of the twentieth century the breed had nearly died out, but a pure line was found in Kenya, taken there by Lady Violet Carnegie in 1902, and re-imported to give the existing stock a boost.

Hens of this breed make very good broodies; they were apparently particularly favoured by gamekeepers to rear pheasants, but their own chicks can easily chill on wet grass, being so close to the ground. As short legs are a feature of this breed, the occasional chick that develops long legs should not be included in breeding stock.

# Scots Grey

A hardy bird well suited to its native climate

**PURPOSE:** Layer
**CLASS:** Soft feather light
**BROODY:** Occasionally
**NO. EGGS PER YEAR:** Moderate
**COMB TYPE:** Single
**EGG COLOUR:** White
**FEATHER COLOURING:** Cuckoo

Hen

As its name suggests, this breed is found mostly in Scotland, where it has been in existence for over 200 years. It is a hardy bird that thrives in even the harshest climate, with the reputation of maturing quickly – an advantage in a country where the days are short in winter.

The Scots Grey has a fairly upright stance and long legs, supposedly as a result of some gamefowl in its ancestry. It is not an easy bird to tame and will roost up trees if given the chance – this is not a breed that will take kindly to close confinement. Although classed as a non-sitter, hens will occasionally go broody, and when they do they make excellent, protective mothers.

# Sebright

The well-deserved winner of the best-turned-out prize

**PURPOSE:** Ornamental
**CLASS:** True bantam
**BROODY:** Occasionally
**NO. EGGS PER YEAR:** Few
**COMB TYPE:** Rose
**EGG COLOUR:** White to pale cream
**FEATHER COLOURING:** Gold laced, silver laced

Sebrights are true bantams, meaning there is no large equivalent. Their creator, Sir John Sebright, worked on breeding programmes of various animals, including chickens, around the year 1800. What he wanted to produce was a laced bantam and this he certainly did, by crossing Nankin, laced Polish and Hamburgh and possibly other breeds, the result being the very smart and beautifully laced gold and silver Sebrights. It is one of the smallest bantam breeds, weighing as little as 550g (19.5oz).

Sebrights are smart little birds with the tail carried high and spread and the head and neck carried well back, causing the breast to be thrust forward. One odd feature of this breed is that the cockerels are 'hen-feathered' and do not have the saddle hackles or sickles of a normally feathered cockerel. This fact is thought to be the cause of the breed's poor fertility and delicate chicks, and a certain amount of skill will be needed for successful breeding.

Gold laced hen

Sebrights are not great egg-layers but they make up for this fact with their charm and character. They make excellent pets, as they are easy to tame, and the cocks have the reputation of being quiet crowers, which can certainly be an advantage if they are to be kept in a built-up area.

# Silkie

This is the powder-puff of the poultry world

Hen

Hen

**PURPOSE:** Layer
**CLASS:** Soft feather light
**BROODY:** Yes
**NO. EGGS PER YEAR:** Few
**COMB TYPE:** Walnut or cushion
**EGG COLOUR:** Pale cream or tinted
**FEATHER COLOURING:** Many varieties

Day-old chick

This is a very ancient breed, originally from Asia, though its exact origins are unknown. Marco Polo brought back stories of these strange fowl when he returned from his travels in the thirteenth century.

Silkies are totally different from most other poultry, having a melanotic gene, which results in their skin, and even their bones, being black. Their very odd feathers are fluffy and more like fur, but do not retain heat or provide waterproofing in the same way as normal feathers, and this causes them to dislike getting wet. The Silkie is also one of the few breeds with a fifth toe and completes its unique look with a topknot and sometimes even a muff.

They make excellent pets, being docile and friendly, but do spend a good deal of their time broody and are widely used as surrogate mothers for all kinds of other birds. Although small, they are not considered to be bantams but classed as soft feather light.

Dark hen

# Spanish

An unusual but elegant choice with a unique charm

**PURPOSE:** Layer
**CLASS:** Soft feather light
**BROODY:** No
**NO. EGGS PER YEAR:** Moderate
**COMB TYPE:** Single
**EGG COLOUR:** White
**FEATHER COLOURING:** Black

The white-faced Spanish is one of the oldest of the Mediterranean breeds, having been known since the sixteenth century, or even before. It is also one of the oldest known in the United States, having been brought there via the Caribbean by early Spanish settlers, and made an appearance at the first poultry show in Boston in 1849.

In the hen the comb droops to one side, although it should remain upright in the cocks, whose large white faces give the birds a rather haughty look, though this can give them an eccentric sort of charm. They are, in fact, rather noisy and not always friendly creatures, which do not take kindly to being over-contained, preferring free range. This breed is rather slow to mature, the white face taking a year or so to develop, but it will eventually reward you with a good number of large white eggs.

# Sussex

An attractive, popular and reliable breed

Light cock

**PURPOSE:** Layer, table
**CLASS:** Soft feather heavy
**BROODY:** Yes
**NO. EGGS PER YEAR:** Very prolific
**COMB TYPE:** Single
**EGG COLOUR:** Light brown
**FEATHER COLOURING:** Black, buff, light, silver

This is one of Great Britain's oldest breeds and also one of the commonest – the light, with its black neck and tail and white body, being the most popular. Originally there was a black-red (now called brown) version, which was well camouflaged when sitting, but a 'light', really Columbian, was developed, as the white feather stubs didn't show on the carcass when plucked. The light was also useful for sex-linkage; typically, a light Sussex hen mated with a Rhode Island Red cock produced chicks with dark females and pale males.

The speckled Sussex has mahogany feathers with black-and-white mottling that glows green in the sun, and there is also a buff-and-silver version. All Sussex hens are hardy, adaptable and easily handled. If you are looking for a pet, then this is the breed for you, as the Sussex has a reputation of being an exceptionally calm bird. Even cocks will tolerate each other. They go broody, make excellent mothers and lay well – in short, the perfect backyard bird.

*Continues overleaf*

Speckled hen
Light hen

# Transylvanian Naked Neck

A conversation-piece breed that is also a reliable layer

**PURPOSE:** Layer, table
**CLASS:** Soft feather heavy
**BROODY:** Occasionally
**NO. EGGS PER YEAR:** Prolific
**COMB TYPE:** Single
**EGG COLOUR:** Tinted
**FEATHER COLOURING:** Black, blue, buff, cuckoo, red, white

Transylvania is now a region of Romania and not necessarily where this breed originated, as they are found in many other parts of Europe and the Middle East. Also known as a Turken because it faintly resembles a turkey, the Naked Neck has to be one of the strangest, some might even say ugliest, breeds of poultry. The lack of feathers on the neck is caused by a gene that reduces the size and density of the plumage. This is an advantage in a hot climate (although they can suffer sunburn on their naked necks), but these mild-mannered, friendly birds are hardy and tolerant of cold. They are also thought to be exceptionally immune to disease.

The fact that they are slow to mature also means that they produce good meaty carcasses, which are quick to pluck owing to the thinness of the plumage. This lack of plumage also has the advantage of putting protein into egg production rather than putting it towards producing feathers, and they are good, reliable layers of tinted eggs.

# Welsummer

An all-round beautiful breed, ideal for the backyarder

**PURPOSE:** Layer, table
**CLASS:** Soft feather light
**BROODY:** Occasionally
**NO. EGGS PER YEAR:** Prolific
**COMB TYPE:** Single
**EGG COLOUR:** Dark brown speckled
**FEATHER COLOURING:** Partridge, silver duckwing, white

This is a fairly modern breed, created by a farmer in Welsum in the Netherlands at the end of the nineteenth century, who crossed various breeds, including Barnevelders, until he found the stable cross he wanted. It didn't reach British shores until the late 1920s and is still not common in the United States.

Apart from being an exceptionally pretty hen and handsome cockerel of a traditional farmyard type, its chief claim to fame is the wonderful dark, sometimes speckled, brown of its eggs. This hardy breed is generally adaptable; it is economical to feed, being an excellent forager, though occasional strains do tend to be on the flighty side. The hen has lovely partridge colouring, enhanced by the light shafts of the feathers. Although classed as 'light', these birds can actually be heavier than some 'heavy' breeds.

Hens will go broody but not excessively so, making reliable mothers, and when chicks hatch an experienced eye can distinguish hens from cockerels by the colour of the stripe on their heads – the cockerel being darker. This does take practice and needs comparison, but by six weeks or so the cockerels will become obvious as their combs start to appear.

# Wyandotte

A superb and showy dual-purpose bird with a strong character to match

**PURPOSE:** Layer, table
**CLASS:** Soft feather heavy
**BROODY:** Yes
**NO. EGGS PER YEAR:** Very prolific
**COMB TYPE:** Rose
**EGG COLOUR:** Tinted
**FEATHER COLOURING:** Wide variety including laced and pencilled

This bird acquired its name from a Native American tribe called the Wyandot or Wendat, which became corrupted to Wyandotte by settlers when this poultry breed was developed in the late nineteenth century. It was designed to be the perfect dual-purpose bird, being a prolific layer and prized for its yellow-skinned meat.

Wyandottes come in a wide range of colours, the silver laced being the original and perhaps most striking variety. They have a rather rotund appearance, with a short tail, sturdy legs and slightly loose feathers, which makes them look fluffy and larger than they actually are.

This is a robust breed, the hens of which make excellent mothers. Most birds are calm and docile, but their strong characters can sometimes make the occasional one seem unfriendly – this is, however, an ideal beginners' breed.

Silver
laced hen

# Rare breeds

## Campine

This breed was developed in Belgium in the district of La Campine from Turkish fowl. They are striking birds that mature early; the hens begin to lay at 18 weeks. They are vigorous and need plenty of space, tending to stay wild, so they will require good fencing or clipped wings if they are to live in a pen.

**PURPOSE:** Layer | **CLASS:** Soft feather light | **NO. EGGS PER YEAR:** Prolific

## Dandarawi

An ancient Egyptian breed from the city of Dendera, north of Luxor. They are tough little birds that can easily look after themselves, enjoy the heat and fly well. This is not a breed that will take kindly to close confinement; however, they are perfect for a free-range life in a farmyard.

**PURPOSE:** Layer | **CLASS:** Soft feather light | **NO. EGGS PER YEAR:** Moderate

## Fayoumi

Known since the time of the pharaohs, this ancient Egyptian breed arrived in Britain in 1984. It is a small, hardy breed with an upright tail, which likes the heat and matures early. Being known for their wild streak, the birds are rather flighty, but could support themselves if allowed to free range. They also seem to be resistant to viral disease, possibly including avian flu.

**PURPOSE:** Layer | **CLASS:** Soft feather light | **NO. EGGS PER YEAR:** Prolific

## Kraienköppe

This breed arose from the crossing of country fowl from the region of Twente on the Dutch–German border with Malays and Leghorns. It will tolerate confinement but is an active forager and competent flyer. The cock birds have exceptionally loud voices that may not endear them to near neighbours.

**PURPOSE:** Layer | **CLASS:** Soft feather light | **NO. EGGS PER YEAR:** Prolific

## Orloff

These rather wild-looking birds originally came from Persian stock but were developed in Russia in the early nineteenth century. They are rather slow to mature into tall, hardy birds with distinctive beards and muffs. Although they lay well in their first year, egg numbers tail off as they age and they would usually be kept for meat.

**PURPOSE:** Ornamental, table | **CLASS:** Soft feather heavy | **NO. EGGS PER YEAR:** Few

## Penedesenca

Although still very rare, this breed has been revived in the Catalan region of Spain since the 1980s. These are nervous birds with an excellent foraging ability, which could survive in the wild if necessary. The eggs are the darkest of purplish browns, in some cases almost black in the first year, gradually becoming lighter with age.

**PURPOSE:** Layer | **CLASS:** Soft feather light | **NO. EGGS PER YEAR:** Prolific

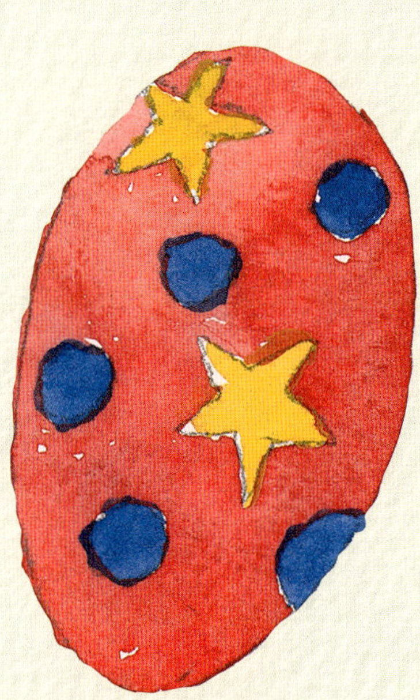

# Crafts

It is not just fresh eggs your flock will provide you with. In this section, you'll find ideas for using eggs and feathers in creative ways. But beware – you may soon find you are scouring the hen run for feathers and keeping empty eggshells to turn into creations of your own!

## Egg crafts

Blown eggs are simply the shell of eggs without their contents. These can be used for many different projects. Although brittle, they last indefinitely.

### HOW TO BLOW EGGS

Using a sharp tool, such as a compass point or darning needle, pierce a hole in the narrow end of the egg. Then use your tool to pierce another hole at the other end of the egg. This hole will have to be a bit larger. Make the hole as neat as you can but don't worry too much, since this end won't be visible when you display your eggs.

Stick your needle or a cocktail stick through the larger hole and wiggle it around inside the shell to break the yolk. Holding the egg over a bowl, blow through the smaller hole. Blow hard, and gradually the broken yolk should come out through the larger hole. If you don't see any yolk emerging, have another go at wiggling the cocktail stick inside to break it up more.

If you really cannot get any egg to come out, you will need to enlarge the larger hole a little, but with practice, you will be able to empty the egg through a remarkably small hole.

### Colour creation

Many colours of dye can be created from everyday household products. To create the following colours, add the dye ingredients below to the water-vinegar mixture (see p. 114):

- **Yellow:** 3–5 tbsp of turmeric
- **Orange:** 10 large red onion skins (you will need to strain out the skins from the dye solution before adding the eggs)
- **Brown:** 30g (1oz) instant coffee granules
- **Pink:** 400g (14oz) chopped beetroot
- **Blue:** 100g (3.5oz) red cabbage (you will need to strain out the cabbage from the dye solution before adding the eggs)

Making natural dyes is trial and error, and your results will vary, but the more ingredients you add to the dye solution, the stronger the colour will be.

Rinse the inside of your shell with water – a syringe will help. Shake the shell with the water inside and then blow out the water through the hole you've made. Allow the shell to dry out completely.

Once dried, the hollowed egg is ready to be decorated or made into a cactus.

## NATURALLY DYED EGGS

There is something delightful about creating your own dye and using it to decorate eggs. The eggs you dye can be raw, hard-boiled or blown (see How to blow eggs, p.113). Dyeing works best on white eggs, but you can achieve great results with brown eggs as well.

To create a dye solution, add your dye ingredients, 500ml (2 cups) water and 2 tbsp vinegar to a saucepan, and boil for a good 30 minutes. The acid in the vinegar creates a chemical reaction that will help the colour bond with the eggshell when the eggs are added to the liquid later. Lemon or lime juice will also provide the same reaction.

Once you have made the dye solution, place the eggs in the pan. To make hard-boiled eggs, boil them in the solution for only 10 minutes; for fresh or blown eggs, let the solution cool completely and leave the eggs in it overnight.

The longer you leave your egg in the solution, the darker the shell will become. If you find that the colour dulls when the eggs are dry, you can rub them with cooking oil or, if they are blown and to be kept, apply a light coat of varnish.

### *Patterned eggs*

For more detailed designs when dyeing eggs, you can print patterns on the eggs using plant leaves – tiny bits of fern, clover leaves, almost anything you can find.

First, attach your patterns to the egg. You will need to tie any leaves into pockets of chopped up nylon tights or muslin to keep them in place – this is a fiddly job but I found that wetting the leaves made them stick to the egg just enough to tuck them into the stretchy pockets. You can also wrap the eggs in wool, or even elastic bands, to produce stripes. The possibilities are endless.

Once your pattern materials are in position, place the eggs in the water-vinegar solution with your dye ingredients and follow the steps above. When the eggs have been in the solution for the appropriate amount of time, unwrap them and reveal your designs.

*Cactus eggs*

Simply paint your egg green, draw on some spines and stick a little fluff of red wool or crumpled-up tissue paper onto the pointed end of the egg to form a flower. Plant your cactus egg in a small flowerpot or cut the cardboard inside of a toilet roll in half and paint it to resemble a flowerpot.

*More ideas for crafting with blown eggs*

- Paint the egg black, then add a design using gold paint or pen.
- You can achieve a batik-inspired look by painting part of the egg black and then scratching a design into the paint (my school compass comes in handy for this).
- Decoupage – cut bits out of magazines or chop up coloured tissue paper and stick them on. Varnish your eggs when they're dry.
- Make Humpty Dumpty eggs – you'll need to buy some tiny craft eyes for this one.

# Eggshell masterpiece

When you're cooking with eggs from your hens, don't throw your eggshells away. Why not make them into unusual 3D artworks? When you break an egg open, keep both halves – rinse them out and put them somewhere warm to completely dry out.

## You will need

6 half shells, or as many as you like
Watercolour and/or gold paint of any sort
Glue
A box frame suitable for displaying 3D items
Small paint brush

## Instructions

1. Paint the inside of your shell either with a single colour, one colour and a bit of gold paint to add a touch of bling, or many colours of your choice. If one shell is larger than the others, carefully break off little bits all the way round until your half shells are all a similar size.
2. Measure out where your shells will sit in the frame and mark with a pencil.
3. Squeeze a blob of quick-drying glue onto each of your marked spots on the base of your frame and place the shell on it.
4. And as simply as that, you will have created an original artwork and conversation piece.

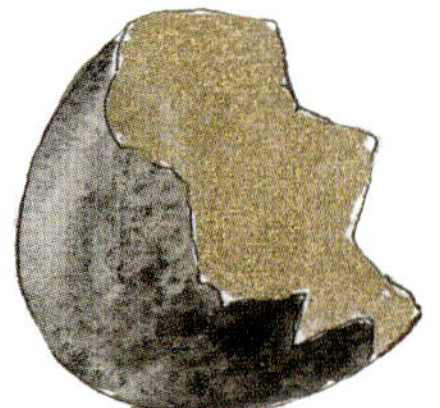

# Eggshell toadstools

These are fun for children to make. There's no need to restrict yourself to a miniature flowerpot; why not create a whole woodland scene?

### You will need

Discarded eggshells that have been cracked in half
Scissors
Red watercolour paint
Small paint brush
White paper discs from a hole punch
Glue
15cm (6in) squares of paper
Miniature flowerpot
A handful of soil or some scrap paper
Moss

### Instructions

1. Wash out the inside of some discarded eggshells and allow the pieces to dry completely.
2. Carefully snip round the edges of the shells with scissors to give them a neat edge.
3. Paint the outside of the shells red, then stick on the little discs of paper you've made with a hole punch.
4. To form stalks for your toadstools, roll up the squares of paper, gluing the paper along the inside edge to create a thin tube. Snip one end of your paper stalk to make it splay out, apply a little more glue to the splayed sections, and stick the glued, splayed end of your stalk to the inside of the painted shell.
5. Fill your flowerpot with a little soil or shredded scrap paper – an old newspaper or junk mail or whatever comes to hand is fine – and plant the pointy end of your toadstool stalk into it.
6. Tuck a little moss around the base of your toadstool to give it a realistic finish.

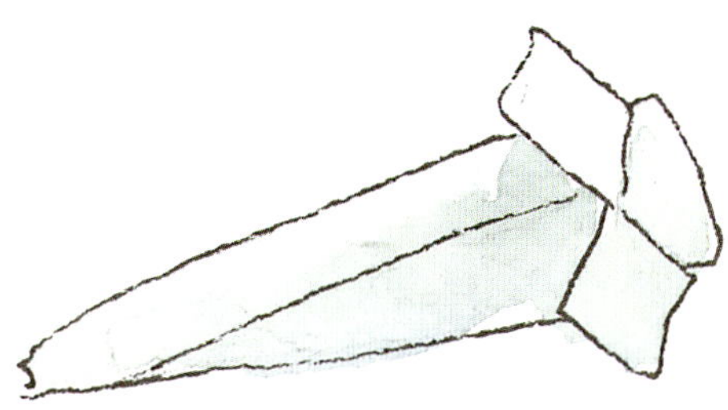

# Mustard and cress heads

Something different to decorate your Easter table, or just for fun. Next time you have soft-boiled eggs for breakfast, save the shells and wash them out, setting them aside until you have the number you want.

### You will need

Discarded eggshells that have been cracked more or less in half
Pen
Egg cup or toilet roll
Cotton wool, kitchen roll or toilet paper
Cress or mustard seed

### Instructions

1. Holding the eggshell with its open side at the top, draw a different face on the side of each of your shells so the face is just below the open top. Set each of your egg faces into an egg cup. If you don't have enough egg cups, you can also cut the cardboard centre of a toilet roll into short tubes and set your egg faces on top of those.
2. Fill the inside of each eggshell with cotton wool, kitchen roll or even toilet paper, then soak this material thoroughly.
3. Sprinkle cress or mustard seed onto the soaked material inside each egg and add a little more water. It is important to keep the water topped up daily.
4. The seeds will germinate in a few days and very soon you will see your eggs' 'hair' begin to grow. When their 'hair' has grown long enough for you to do a little hairdressing, you can trim the first growth, and a second will appear.

# Feather crafts

When your hens moult, they will drop feathers all over the run – pick them up and store them carefully; you never know when they may come in useful! Even if one dies, although this may seem gruesome, why waste the beautiful feathers?

To make absolutely sure there are no mites in the feathers just pop them in the freezer for a week or so, then store them in an airtight container.

### PLACE NAME CARDS

A simple but effective use of wing or tail feathers is to use them to decorate place settings at a table. Is it a special occasion? A gold or silver wedding? Simply dip the feather tips in gold or silver paint or spray with photo mount or any aerosol glue and sprinkle with glitter, before attaching them to place cards.

If you're lucky enough to have a hen with spangled or barred feathers then they are beautiful on their own.

# Feather earrings

For this you will need what are called 'findings', essentially the parts of the earring that attach to the wearer's earlobe. These come in all shapes and sizes, and once you've discovered just how easy this is you may find you are making a great many – wonderful small gifts that cost virtually nothing.

### You will need

- 2 pinch catches, sometimes called calotte clamshells
- 2 ear hooks
- 2 matching feathers, or two or three similar ones for each earring
- Drop of glue
- Small pliers

### Instructions

1. Pull off the fluffy bit at the base of the feather and snip so that there is just enough of the shaft to be enclosed in the pinch catch. Put a drop of glue into just one side of the catch and shut the feathers in.
2. Twist open the top ring of the pinch catch with your small pliers. You can do this with your fingers but it is a bit fiddly.
3. Feed the open end through the ring at the bottom of the ear hook and close. Repeat for the second side. Admire your beautiful earrings!

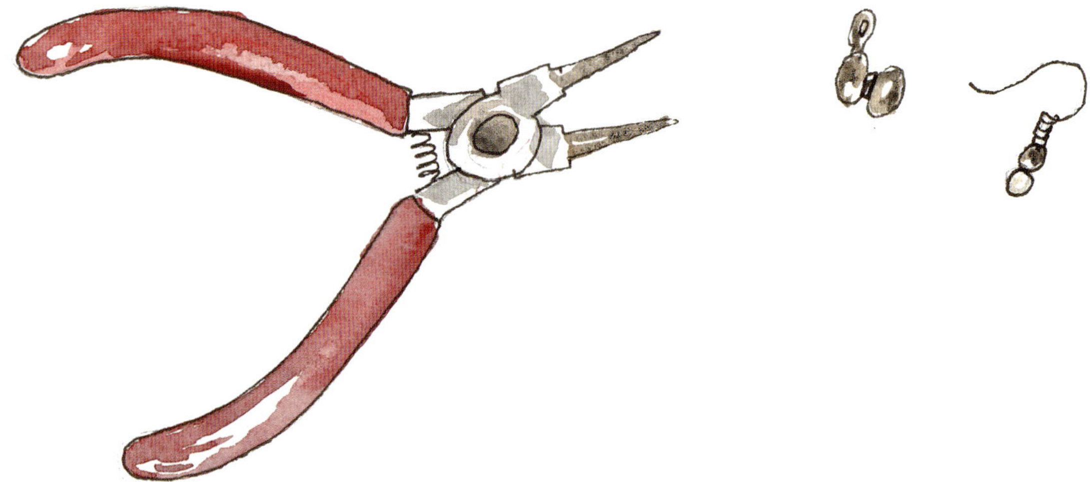

# Feather printed greeting cards

A lovely way of making your own greeting cards is to literally print using feathers collected from your own hens. Ready-made blank cards are easy to obtain and come with envelopes of the correct size, or you can simply fold a sheet of A5 paper in half.

## You will need

- Printing ink or acrylic paint
- Paint tray or palette – this could be an empty meat tray, a piece of cardboard covered with clingfilm or tin foil, or any smooth flat surface
- Printing roller
- Feathers
- Newspaper
- Ready-made cards or A5 paper (folded in half)

## Instructions

1. Squeeze some ink (or paint) onto your palette, and roll the roller in it until the roller is evenly coated with ink (or paint).
2. Place your feathers on a sheet of newspaper and carefully roll the ink over them so that they are now coated.
3. With your card folded in half, very carefully lift the feathers off the newspaper and place them artistically ink side down on your card.
4. Cover with a clean sheet of newspaper and rub gently over the top. Rub with a circular motion rather than backwards and forwards. This will minimise the risk of the feathers slipping and creating a smudged impression.
5. Lift the newspaper and the feathers off gently. You now have a beautiful home-made card to be proud of!

## Alternative Method

If you do not have a roller, do not despair. You can use a brush instead to spread the ink all over the palette, trying to create a smooth and even coating. Place your feathers on the ink-covered palette. Then place the feathers ink-side down onto your card, cover them with a sheet of newspaper and gently rub. Discard the newspaper and feathers.

# Recipes

## Ideas for using up surplus eggs

Can you even imagine life without eggs? What wonderfully versatile things they are in the kitchen. You can boil, scramble and poach them, eat them hot or cold, bake cakes with them, make quiches, tortillas, mousses, soufflés and creamy sauces with them, and so much more.

When you keep chickens, you may find you suddenly have a surplus of eggs – possibly in the spring when your hens are in full lay. You can give your surplus eggs away to friends or family, or try to sell some of them, but you might still find you have more than you can get through, so the pages that follow include a few recipes that can help you make the most of this bounty.

We all have times when we only need to use the yolks or the whites from our eggs. Some of these recipes are a useful way of using up whatever we've been left with in those moments, too.

### EGG SIZES

In the UK, egg sizes are based on weight ranges, while in the US, they're based on minimum weight per dozen. A UK medium egg is comparable to a US medium egg, but a UK large egg is similar to a US extra-large egg, and a US large egg is closer to a UK medium egg.

| UK egg sizes | |
|---|---|
| Small | <53g or <1.9oz |
| Medium | 53–63g or 1.9–2.2oz |
| Large | 63–73g or 2.2–2.6oz |
| Very Large/Extra Large | >73g or >2.6oz |

| US egg sizes (based on minimum weight per dozen) | |
|---|---|
| Medium | 49.6g or 1.75oz |
| Large | 56.8g or 2oz |
| Extra Large | 63.8g or 2.3oz |
| Jumbo | 70.9g or 2.5oz |

## COOKING WITH WHOLE EGGS

You can scramble, poach, fry or hard boil whole eggs, and they can be incorporated into quiches, flans, curds (see pp. 142–143) and crème caramel.

Omelettes are a great way to get through plenty of eggs, and almost anything can be added to your eggs to create a tasty omelette, from cheese, tomatoes, ham and herbs to leftover beans or potatoes.

Making a frittata, which is like a crustless quiche or an omelette that hasn't been folded, will also use up plenty of eggs. You begin cooking frittatas on the stove but instead of folding or flipping it, you finish it off in the oven to set the eggs. Like omelettes, you can fill frittatas with your choice of vegetables, meats and cheeses. See pages 138–139 for a suggested frittata recipe.

Spanish tortillas are another good way to use up lots of eggs. They are traditionally made with eggs, potatoes and onions, and are cooked on the stove and flipped halfway, before being served warm or cold.

Hard-boiled eggs can be pickled or devilled – see pages 132–135 – and they are a vital component of kedgeree, Scotch eggs and salad niçoise.

## COOKING WITH EGG WHITES

Meringues are top of the list for using up egg whites – freshly made meringues are perfect for pavlova (see pp. 148–149) – and once they are made, they will keep in airtight containers for weeks. You can also use egg whites in soufflés (see pp. 140–141) as well as mousses, sorbets, marshmallows and macarons.

## COOKING WITH EGG YOLKS

You can use up egg yolks with delicious sauces (see pp. 130–131) and custards. For a truly extravagant treat, why not try a crème brûlée (see pp. 150–151).

# Quick hollandaise sauce

Serves: 4–6
Time: 15 minutes

Eggs Benedict is a decadently delightful way to start your day, and it's simply a poached egg placed on top of a slice of warmed ham – or bacon – on a halved, toasted muffin with a generous dollop of creamy hollandaise sauce poured over the top. Here's how to make a quick and easy hollandaise sauce with your blender.

## Ingredients

3 egg yolks, at room temperature
125g (4.5oz) unsalted butter
1 tbsp lemon juice
Pinch of cayenne pepper, or to taste
1 tsp Dijon mustard
Salt, a pinch, or to taste
Pepper, 3 or 4 grinds

**Note:** Hollandaise sauce can also be served cold in many other guises; for example, as a sauce to go with asparagus or poached salmon. It will thicken a little more as it cools.

## Method

1 Put the yolks into the jar of a blender.

2 Melt the butter in a small saucepan over a medium heat, then allow it to cool for a minute. Set it aside – it's important to keep it warm.

3 Process the egg yolks with your blender for a couple of seconds, then continue to run the blender as you slowly and carefully add in the still-warm, melted butter until it's incorporated and the mixture has thickened. Add the lemon juice, cayenne, Dijon mustard and seasonings to taste and blend everything again to combine. If your sauce gets too thick, add a splash of water. Pour your sauce into a bowl and serve while it's still warm.

Variation

# Bearnaise sauce

Serves: 4–6
Time: 15 minutes

This daughter of hollandaise sauce is a classic accompaniment drizzled on top of fish or steak, and it tastes great poured over roasted vegetables. The addition of tarragon and infused vinegar makes Bearnaise piquant and creamily rich.

This Bearnaise recipe is also quick and easy to make with a blender, and like its parent sauce, Bearnaise combines well with a freshly poached egg, giving you two opportunities to use up surplus eggs in a single dish.

## Ingredients

**For the infused vinegar:**

60ml (¼ cup) dry white wine
60ml (¼ cup) white wine vinegar
1 small shallot, peeled and very finely sliced

**For the sauce:**

3 egg yolks, at room temperature
Salt, a pinch, or to taste
1 tsp lemon juice
1½ tbsp infused vinegar
125g (4.5oz) unsalted butter
2 tbsp tarragon, finely chopped
Pepper, 3 or 4 grinds

## Method

1. Make your infused vinegar first. Put the white wine, white wine vinegar and sliced shallot in a saucepan and bring to the boil. Lower the heat and simmer the liquid until it has reduced to about 1½ tbsp. This should take about 5 minutes, but keep an eye on it as it can happen very suddenly. Remove from the stove and allow the liquid to cool.
2. Put the egg yolks in a blender, season with salt, add the lemon juice and blend to combine, then add the infused vinegar.
3. Melt the butter in a small saucepan over medium heat and bring to a simmer – the melted butter needs to be hot enough to cook the eggs when they are combined in the next step. Using a saucepan with a pouring spout or transferring the melted butter to a jug will make it easier to pour the butter slowly into the blender.
4. Turn on the blender, and very slowly pour the hot butter into the seasoned egg mixture. When the butter is combined and the sauce is smooth and thick, stir in the chopped tarragon leaves. Check the seasoning and serve.

**Note:** Sauce Paloise is a more recent variation of Bearnaise sauce. To make it, follow the steps for making Bearnaise sauce above and simply replace the tarragon with finely chopped mint.

# Devilled eggs

Serves: 6–12
Time: 10 minutes

Devilled eggs are an easy way of using whole eggs and creating either a first course or a canapé. As you are adding other ingredients to the yolks when you make the filling, you will probably have more than you require when you re-fill the eggs – but the filling will be just as delicious spread on little toasts or blinis.

For all the below fillings, you first need to hard boil your eggs, and the ingredient quantities below are to combine with 6 eggs.

### Method

For each recipe, do the following:

1. First, boil your 6 eggs for 10 minutes, then plunge them into cold water to stop them cooking before peeling them. Note: very fresh eggs may be difficult to peel, the shell breaking off in small pieces. When hard boiling, use eggs that are at least a week old.
2. Slice the peeled eggs in half lengthwise, spoon out the yolks, put them into a bowl and mash them together, then combine your mashed yolks with one of the options opposite to create your filling.
3. Finally, use a piping bag to re-fill the whites of your eggs with your mixture.

## Classic devilled eggs

To the mashed yolks, add 1 tsp Dijon mustard, 2 tbsp mayonnaise, 1 tsp (or to taste) of Tabasco sauce and 1 tbsp of very finely chopped onion, combining the ingredients well to create your filling. When you have refilled the eggs, sprinkle them with paprika.

## Devilled avocado eggs

To the mashed yolks, add 1 avocado (peeled, stoned and mashed), 1 tbsp lime (or lemon) juice, 1 tbsp sour cream, 1 jalapeno pepper finely sliced (leave the seeds in if you like it hot or remove), combining the ingredients well to create your filling. Sprinkle the filled eggs with finely chopped chives.

## Smoked salmon eggs

To the mashed yolks, add 100g (3.5oz) minced smoked salmon, 1 tbsp lemon juice and 2 tbsp sour cream, combining the ingredients well to create your filling. Grind pepper and a sprinkle of chopped parsley over the filled eggs.

## Devilled blue cheese eggs

To the mashed yolks, add 2 tbsp mayonnaise, 2 tbsp blue cheese, 2 tbsp finely chopped celery, 1 tsp chilli sauce (or to taste), combining the ingredients well to create your filling. Sprinkle the filled eggs with paprika.

# Pickled eggs

Serves: up to 6
Time: 15 minutes

One way of saving excess eggs is to pickle them. Placed in airtight jars in pickling vinegar, eggs will keep for up to 3 months, or longer in the fridge. If you fancy pink pickled eggs, add beetroot juice to the pickling recipe, or turmeric to turn them yellow. Over time the colour will penetrate right through the whites.

## Ingredients

6 eggs

**For the pickling liquid:**

400ml (1¾ cups) white wine vinegar
100g (3.5oz) caster sugar
2 bay leaves
1 tbsp black peppercorns
1 tbsp coriander seeds
1 tbsp mustard seeds
½ tbsp salt

## Special equipment

Jars with lids, sterilised and dried

## Method

1. Make the pickling liquid first. Put all the ingredients in a saucepan over a medium heat and bring to the boil. Boil for 2 minutes, then allow the liquid to cool. If preferred, you can sieve out the spices at this stage, but the flavour will be enhanced if you leave them in.
2. When your pickling liquid is sufficiently cool, boil 6 eggs for 8 or 10 minutes, depending on their size, then plunge them immediately into cold water – this will prevent the yolks acquiring a black edge.
3. Peel the eggs and put them into a dry, sterilised jar with an airtight lid – cover the eggs completely with the pickling liquid.
4. Leave for a minimum of two weeks (ideally a month) before eating, and store in the refrigerator once opened. Use within three months.

**Note:** You can add any other herbs or spices you fancy such as chilli flakes, fennel seeds, allspice or buy a bag of dried pickling spice, which often come with all the spices included.

## Sterilising jars

Wash the jars or run them through the dishwasher. Put the jars on a baking tray in the oven heated to 140°C (275°F)/gas mark 1 for about 10 minutes. Allow to cool before filling.

# Shakshuka

Serves: 3–4
Time: 20 minutes

Originally from North Africa, this is an easy and tasty dish that's perfect for a quick lunch or supper.

### Ingredients

- 1 tbsp olive oil
- 2 red onions, peeled and chopped
- 2 red chillies deseeded and finely sliced
- 2 garlic cloves, finely sliced
- 2 cans cherry tomatoes (or chopped plum tomatoes)
- 2 tsp sugar
- 6 eggs
- Bunch of coriander and any other herb you have, such as parsley or thyme, finely chopped

### Method

1. Heat the oil in a frying pan that has a lid (if you do not have one, do not despair, you can create a lid for any frying pan with a double layer of tin foil).
2. Add the onions, chilli and garlic to the pan and fry them gently on a medium heat for about 5 minutes, or until they are soft.
3. Stir in the tomatoes, sugar and half the herbs, and simmer for about 10 minutes until the sauce has slightly thickened.
4. With the back of a large spoon, create 6 indentations in the sauce and crack an egg into each one. Put a lid on the pan and continue to gently simmer until the eggs are lightly cooked. This usually takes about 5 minutes.
5. Sprinkle over the remaining herbs and serve in the pan with crusty bread.

# Asparagus and feta frittata traybake

Serves: 6–8
Time: 30 minutes

A perfect lunch for a large family, this is delicious when served with a crunchy mixed salad.

## Ingredients

250g (9oz) asparagus, woody ends removed
12 eggs
250ml (1 cup) milk
2 tsp Dijon mustard
2 good handfuls baby spinach
75g (2.5oz) feta (or crumbly goat's cheese, if preferred)
6 or 7 grinds of pepper

## Method

1. Heat the oven to 190°C (375°F)/gas mark 5.
2. Lightly oil a baking tray (with sides a minimum of 2cm (1in) high).
3. Saving the top 8 or 10cm (3 or 4in) of the asparagus on one side, slice the rest diagonally into 3cm (1.5in) slices; halve the slices if the asparagus is thick.
4. (I like to blanch the asparagus at this point for about 60 seconds, but if you like a bit of crunch, you can go straight on to the next stage.)
5. In a large bowl, whisk the eggs, milk, mustard and pepper. Stir in the spinach and sliced asparagus and pour onto the prepared baking tray. Arrange the asparagus spears in the egg mixture, then crumble over the feta.
6. Put the sheet in the oven (middle shelf) and bake the frittata for approximately 15 to 20 minutes, or until it has puffed up and the eggs are no longer wobbly.
7. Remove it from the oven and allow it to rest for a few minutes before serving with salad. The frittata will be just as delicious cold.

# Cheese soufflé

Serves: 6–8
Time: 25 minutes

This is something you may be able to throw together at the last minute, as you will probably have the ingredients in your larder or fridge. Although the recipe suggests using Gruyère cheese, any hard cheese would work. With soufflés, the tricky part is the timing, and a little practice may be needed to get your soufflés spot on.

## Ingredients

150ml (½ cup) milk
25g (1oz) plain flour
25g (1oz) unsalted butter
1 tsp Dijon mustard
4 eggs – use 4 whites and 2 yolks
85g (3oz) grated Gruyère cheese (or hard cheese of your choice)
Salt and pepper
Pinch of cayenne pepper to finish

## Special equipment

6 ramekins, buttered

## Method

1. Preheat the oven to 200°C (400°F)/gas mark 6.
2. Put the milk in a saucepan over a medium heat and heat it until it is almost boiling.
3. In a separate saucepan, melt the butter, stir in the flour and cook for 1 minute to make a roux. Gradually beat in the hot milk and cook until you have a thick sauce. Stir in the mustard.
4. Allow the sauce to cool a little, then beat in the yolks and most of the cheese.
5. While your sauce is cooling, whisk the egg whites in a clean, dry bowl until you have soft peaks. Gently fold them into the sauce.
6. Divide the mixture between the ramekins. Running your finger around the edge of the ramekins will give your soufflés a neat edge when they rise.
7. Sprinkle on the remaining cheese and a pinch of cayenne pepper.
8. Place the ramekins in the oven, on a baking sheet, and bake for around 15–17 minutes, or until they've risen and look golden. Don't open the oven while the soufflés cook as the temperature change could make them collapse.
9. Serve immediately.

LEMON
CURD

# Lemon curd

Makes: 3–4 jars
Time: 20 minutes

Lemon curd is a simple method for using up surplus eggs in a way that lets you enjoy them for a bit longer. It's delicious on toast or even as a sponge cake filling.

## Ingredients

4 lemons, zest and juice
200g (7oz) caster sugar
100g (3.5oz) butter
4 eggs, beaten

## Special equipment

Jars with lids, sterilised and dried

## Method

1. Juice and zest the lemons, then place the lemon juice, zest, caster sugar and butter in a bain-marie or heatproof bowl over a pan of simmering water. Stir until the butter has melted. Using a whisk or fork, stir in the beaten eggs. Keep whisking over the heat until the mixture has the thickness of custard. This may take anything from 10 to 20 minutes.
2. Pour the curd mixture into sterilised jars and set aside to cool. If you prefer not to include the zest, pour the mixture carefully through a sieve first, but if the zest is left in, it will enhance the curd's flavour even more in the jar.

**Note:** Lemon curd is best kept in the fridge, where it will last 2 weeks if stored in an airtight container.

# Lemon pudding

Serves: 6
Time: 20 minutes

If you need a pudding at the last minute, this is the recipe for you. It's made from ingredients you probably already have, and it couldn't be simpler.

### Ingredients

45g (2oz) butter
3 lemons, with rinds grated
60g (2oz) plain flour
225g (8oz) caster sugar
2 large eggs, separated

### Method

1 Put 550ml (1 pint) of water with the butter and the grated rind of two lemons into a saucepan over a medium heat, and bring to the boil.

2 Mix the flour and sugar together in a bowl. Make a well in the centre and pour in the hot liquid, whisking all the time to avoid lumps forming.

3 When everything is combined, return a little of the mixture to your saucepan and whisk in the egg yolks. Stir in the remainder of the mixture and bring slowly to the boil. Cook gently for 10 minutes and allow to cool. Stir in the juice of two lemons.

4 Meanwhile, whisk the egg whites until they are stiff. Fold them into the lemon mixture. Pour into a serving bowl and chill until set.

5 Decorate with slices of the remaining lemon.

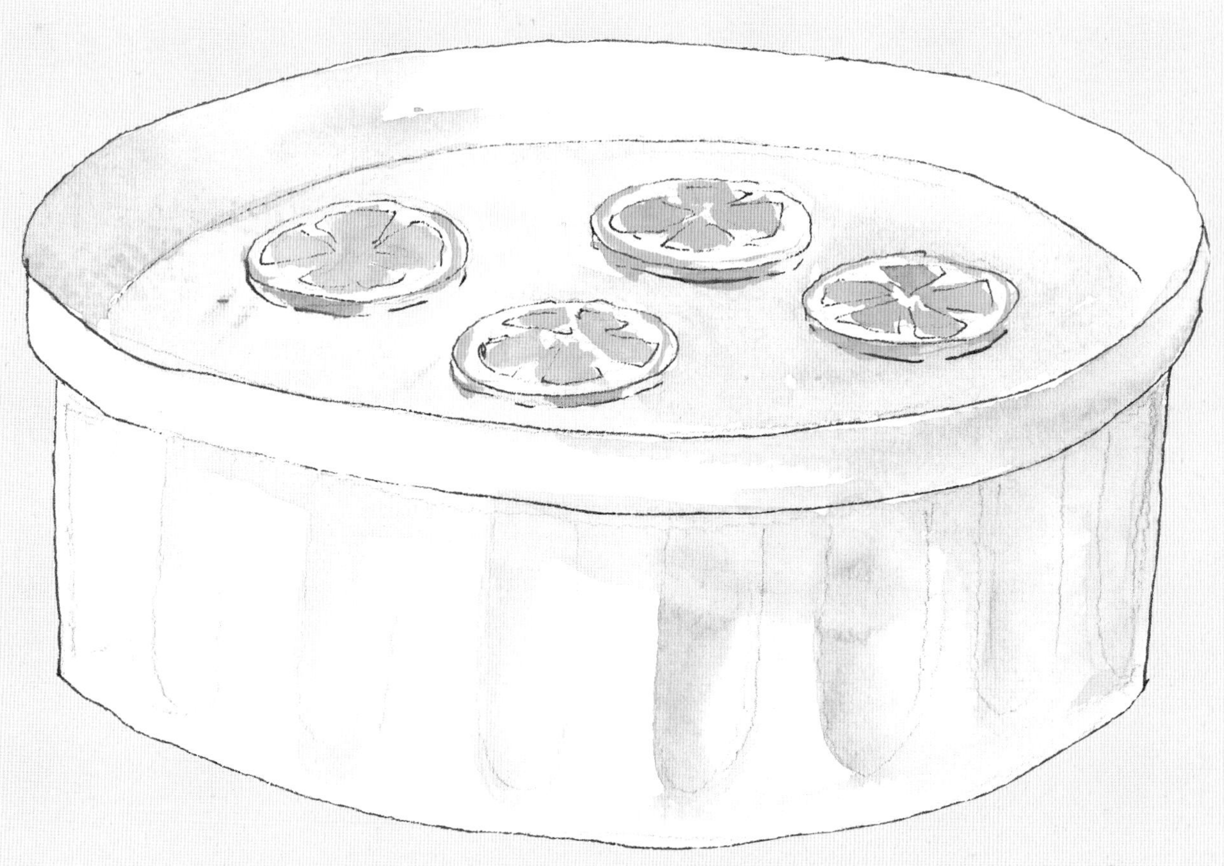

# Oeufs à la neige or îles flottantes

Serves: 4
Time: 40 minutes

Also known as floating islands, this recipe has centuries-old roots. It first appears in recipe books dating from the seventeenth century in different forms and under different monikers. It was eventually named 'floating islands' or 'îles flottantes' by French chef Auguste Escoffier, who published his recipe for the dessert in 1902. This is a simplified version of the French classic.

## Ingredients

4 eggs, separated
Pinch of salt
150g (5oz) caster sugar
500ml (2 cups) whole milk
1 vanilla pod (or 5 drops vanilla essence)
Flaked almonds to decorate

## Method

1. Put the egg whites in a bowl with the salt and 50g (2oz) of the sugar and beat until stiff.
2. Put 375ml (1½ cups) of the milk in a saucepan with the vanilla pod and bring to the boil. Reduce the heat to bring the milk to a simmer.
3. Spoon the egg whites into four oval shapes and gently place them in the milk to poach for 2 or 3 minutes. Carefully lift the poached egg whites out of the milk with a slotted spoon and leave to drain.

### To make the custard

1. Put the remaining 125ml (½ cup) of milk and the poaching milk into a fresh saucepan, add the remaining 100g (3.5oz) of sugar into the liquid and heat almost to boiling.
2. Whisk the yolks in a bowl and very gradually whisk in the hot milk. Return to the heat and stir continuously until the mixture is creamy and thick and coats the back of a spoon.
3. Leave the custard to cool in the dish you are using to serve and, when it's set, top it with the poached egg whites.
4. Serve warm or cold and garnish with flaked almonds.

# Pavlova

Serves: 6
Time: 1 hour 30 minutes

Pavlova is such an easy yet dramatic-looking dessert, and can be whipped up in no time. Whatever fruit you may have can be used, or a mixture of two or three. Try strawberries, raspberries, peaches, mango, kiwis or blueberries – the sky's the limit.

## Ingredients

4 large egg whites
250g (9oz) caster sugar
300ml (1¼ cups) double cream
Fruit of your choice to fill

## Method

1 Preheat the oven to 160°C (320°F)/gas mark 3. Lay a sheet of non-stick baking parchment or oiled greaseproof paper on a baking sheet.

2 Put the egg whites into a large bowl and whisk until stiff. Add the sugar a teaspoonful at a time, whisking well after each addition, until all the sugar has been added. Spread the meringue mixture to form a circle on the baking sheet, building up the sides so they are higher than the middle. Use a piping bag for a professional look or simply spoon and smooth it.

3 Place the sheet in the oven and immediately reduce the temperature to 150°C (300°F)/gas mark 2. Bake for about 1 hour until the meringue is firm to the touch and a pale beige colour. Turn the oven off and allow the meringue to become quite cold while still in the oven. If you keep the oven door closed, you will encourage a more marshmallowy meringue.

4 Remove the cold meringue from the baking sheet and parchment and slide onto a serving plate. Whip the cream and use it to top the meringue, add the fruit of your choice, then chill in the fridge for 1 hour before serving.

**Note:** The meringue can be prepared ahead of time and will keep for several days in an airtight container. Add the fruit and cream close to when it will be served.

# Crème brûlée

Serves: 4
Time: 50 minutes

An indulgent, if rich, dessert – a chef's blowtorch is the best way to create the crunchy caramel topping.

## Ingredients

300ml (1½ cups) double cream
100ml (½ cup) milk
1 vanilla pod
5 egg yolks
50g (1.75oz) caster sugar, plus extra to create the crunchy topping

## Special equipment

4 ramekins

## Method

1. Heat the oven to 180°C (350°F)/gas mark 4.
2. Place 4 ramekins in a roasting tin or cake tin that has enough hot water in it to come about 1.5cm (0.6in) up the sides of the ramekins.
3. Pour the cream into a saucepan with the milk.
4. Split the vanilla pod, scrape out all the seeds and add them to the cream mixture – drop the pod in as well.
5. Put the egg yolks and sugar in a bowl and whisk for 1 minute until the combined mixture is pale in colour and fluffy.
6. Place the pan with the cream on a medium heat and bring the cream almost to the boil. As soon as you see bubbles appearing around the edge, take the pan off the heat.
7. Pour the hot cream into the beaten egg yolks, stirring with a whisk. Remove the vanilla pod from the pan and discard.
8. Pour the cream mixture into the ramekins. Cover with a baking sheet or piece of tin foil.
9. Bake for 30–35 minutes until the mixture is just set – it should wobble and not be too firm.
10. Allow the mixture to cool.

At this stage, the crème brûlées can be kept in the fridge overnight if desired. Once you are ready to serve, sprinkle a heaped tsp of caster sugar over each ramekin to cover the top completely. Use a blowtorch or place under a very hot grill to caramelise the sugar and create the crunchy topping.

# Little egg yolk madeleines

Makes: 25–30
Time: 30 minutes

These delicious sweet 'cookies' are simple to make, and the use of egg yolks gives a soft, almost cake-like texture. The lemon juice adds a burst of tangy flavour.

## Ingredients

150g (5oz) butter
100g (3.5oz) sugar, plus more for rolling
3 egg yolks
150g (5oz) plain flour
1 tsp baking powder
1 tsp vanilla essence
1 lemon, zest and juice

## Method

1. Preheat the oven to 175°C (350°F)/gas mark 4. Lay a sheet of non-stick baking parchment or greaseproof paper on a baking tray.
2. Cream the butter, sugar and egg yolks together in a large bowl.
3. Add the flour, baking powder, vanilla essence and lemon juice and zest, and beat until combined.
4. Form the dough into small balls, each about the size of a walnut.
5. Roll the balls in sugar, before placing them on the baking tray and flattening with a fork. Leave a little space between the balls as they will spread during cooking.
6. Bake for 8–10 minutes.
7. Leave to cool before transferring to a wire rack to cool completely.

# Preparing chickens for the table

If you are rearing your chickens for meat, then you must ensure that you choose the right breed – one that is either dual-purpose or a table bird.

The most important thing is to ensure that your birds are killed humanely (see p.29 on culling). If you are not confident that you can attempt this without causing pain, distress or suffering to the bird, then you should get someone more experienced to do it for you. If you are dispatching the bird at home, make sure you are up to date with local legislation. In the UK, you can only supply home-slaughtered chickens to your own household.

When preparing the bird, it is best to ensure that your chicken has not eaten beforehand so its crop is empty. A full crop can lead to spoilt meat. Once the chicken is dead, hang it upside down and drain the blood over a bucket by removing the head. Make sure you prepare the chicken somewhere that you can disinfect thoroughly.

## PLUCKING

It is best to pluck a bird immediately after the blood letting, while it is still warm. The easiest way to do this is by scalding the bird – fill a large cooking pot with not-quite-boiling water (82°C/180°F) and submerge the bird for a minute or two. The softened feathers will then be easy to pull out. Start with the tail and wings (if you are planning to cook without the wings, these can simply be cut off, still feathered). Be careful as the skin will be more delicate as a result of the scalding and will tear if you are too rough.

Alternatively, you can pluck it perfectly well a day or two later without scalding it – store the bird somewhere cool, such as a larder, until it is plucked.

## DRESSING

On a clean and sanitised workspace, slit the skin on the neck, so that you can pull the neck away from the body and cut it off. Pull the remaining crop and windpipe out as much as you can and cut these away too. Now, you need to gut the chicken. Make an incision at the tail end and enlarge the opening. Cut around the vent with kitchen shears, making sure not to puncture the intestines or stomach. Gently pry open the body cavity and pull out the bird's internal organs, including the lungs, giblets (neck, gizzard, liver and heart) and intestines. Thoroughly rinse out the bird to clean off all the remaining guts. Save the giblets to make stock.

If you aren't going to cook the chicken straight away, put it in the fridge overnight and then freeze it the next day.

## How to joint a chicken

1 Pull the bird's leg away from the body and, using a sharp knife, slice down to the thigh joint. Break at the joint and cut away the whole leg. Repeat with the other leg.
2 Separate the drumstick from the thigh.
3 Slice through the outer breast meat towards the wing joint. Sever the wing from the body. Fold the breast meat over the wing joint. Repeat with the other wing.
4 Find the natural division in the rib cage. Slice along it to separate the breast from the lower carcass.
5 Cut the breast meat into serving portions.
6 Use the remains of the carcass to make stock.

# Glossary

**auto-sexing:** a breed in which male chicks are lighter in colour than females

**bantam:** small version of large chicken

**barring:** stripes of two colours across a feather

**beard:** feathers in a small clump under beak, e.g. Faverolles

**booted:** having feathers on the legs and feet, also includes vulture hocks where long feathers extend downwards from the back of the thighs

**cock:** male bird after its first moult

**cockerel:** male bird before its first moult

**comb:** fleshy protuberance on bird's head

**crest:** feathers on top of a bird's head, known as 'topknot' or, in an Old English Game, 'tassel'

**crop:** area at the bottom of a bird's neck where food collects before passing on to the gizzard

**dubbing:** removal of male gamefowl combs and wattles to prevent injury in fights

**earlobes:** area of bare skin below chicken's ear. Colour denotes egg colour, i.e. in most cases white earlobes = white egg, red earlobes = brown or tinted egg

**egg tooth:** a horny growth on the end of the chick's beak, used to break through the shell when hatching

**gizzard:** the part of the chicken's digestive tract that contains grit to grind down the food

**hackles:** long, narrow feathers on bird's neck, also saddle feathers on male

**hard feather:** describes the short, narrow, tight-fitting feathers of a gamefowl

**hen:** female bird after its first moult

**hen-feathered:** a breed where the male bird lacks sickles or hackle feathers, e.g. Sebright

**hock:** the joint on a bird's leg, between the thigh and the lower leg

**keel:** the breastbone, often featherless in gamefowl

**leader:** the backward-pointing spike of a rose comb

**meat spot:** small harmless spot of blood in the egg

**moult:** annual shedding of feathers, during which a hen stops laying

**muff:** feathers that protrude from both sides of the face in combination with a beard

**parson's nose:** lump of flesh from which tail feathers grow (technically known as uropygium)

**point of lay:** hens about to lay their first eggs, from 18 weeks old

**pullet:** female bird before its first moult

**rooster:** US term, in theory used to describe a male after its first moult. In practice, it is used to describe a male chicken

**saddle:** bird's back in front of the tail

**sickles:** long, curved tail feathers of male birds

**soft-feathered:** a breed that has plumage that is looser and fluffier, e.g. Brahma

**spike:** see 'leader'

**spurs:** sharp, horny growths on the legs of male and some female birds; also known as 'cockspurs'

**stag:** male gamefowl before its first moult

**trio:** group of two hens and one cock

**true bantam:** small chicken that has no large equivalent

**uropygium:** see 'parson's nose'

**vent** or **cloaca:** orifice through which eggs and excretions are passed

**vulture hocks:** feathers that grow from the hock joint and point down

**wattles:** fleshy appendages that hang below the beak

**wing clipping:** when primary and secondary feathers are cut off or clipped on one wing to unbalance bird and prevent flight

# Useful websites

**adoptabirdnetwork.com**
Rescue chicken adoption site (US).

**amerpoultryassn.com**
The site of the American Poultry Association which offers support, exhibitions and education for poultry enthusiasts in the US.

**bantamclub.com**
Site of the American Bantam Association, a national organisation that promotes the breeding and exhibiting of bantam poultry.

**bhwt.org.uk**
Hybrid hen rehoming site (UK).

**cdc.gov/healthy-pets/about/backyard-poultry.html**
CDC (US Centers for Disease Control and Prevention) – useful site with US-specific resources, guidance and recommendations for keeping you and your flock disease-free.

**gov.uk/government/organisations/department-for-environment-food-rural-affairs**
Defra (Department for Environment, Food and Rural Affairs) – very comprehensive site providing updates on avian influenza.

**freshstartforhens.co.uk**
Hybrid hen rehoming site (UK).

**poultryclub.org**
The site of the Poultry Club of Great Britain. Gives lists of breed clubs and societies with their secretaries and phone numbers. Also, a great deal of useful advice, from breed lists to care, etc.

**rarepoultrysociety.com**
This society was formed to cater for the rare and endangered breeds that do not have their own breed club or society, and includes new breeds from abroad.

**rbst.org.uk**
The site of the Rare Breeds Survival Trust, a charity founded to protect all native breeds of farm animals, including poultry.

# Index